Diffusion Weighted Imaging
of the Gastrointestinal Tract

Sofia Gourtsoyianni · Nikolaos Papanikolaou
Editors

Diffusion Weighted Imaging of the Gastrointestinal Tract

Techniques and Clinical Applications

 Springer

Editors
Sofia Gourtsoyianni
Department of Radiology
Konstantopouleio General Hospital
Athens
Greece

Nikolaos Papanikolaou
Head of Computational Clinical Imaging
Group
Centre for the Unknown, Champalimaud
Foundation
Lisbon
Portugal

ISBN 978-3-030-06524-9   ISBN 978-3-319-92819-7   (eBook)
https://doi.org/10.1007/978-3-319-92819-7

This Springer imprint is published by the registered company Springer Nature Switzerland AG
The registered company address is: Gewerbestrasse 11, 6330 Cham, Switzerland

Preface

Imaging the GI tract with MRI has been revolutionary especially for patients with inflammatory bowel disease and in particular for the assessment of small bowel and perianal Crohn's disease. In addition MRI is the imaging modality of choice for locoregional staging of rectal cancer. Imaging the GI tract with morphological MR sequences hinders challenges as ideally the bowel needs to be void of content and adequately distended and opacified to avoid misinterpretation of bowel wall thickness, which is not always the case.

Technical advances in gradient performance, radiofrequency coils, and pulse sequences have increased the clinical applications of diffusion weighted imaging (DWI) beyond the central nervous system. Especially in GI tract DWI has been shown to provide an additional contrast mechanism that renders GI pathology more conspicuous.

As radiology is moving from the pure anatomical imaging techniques to functional-molecular imaging, the use of DWI has gained great acceptance. In the era of "imaging safely" DWI acts as an indispensable noninvasive problem solver. Although it does not seem to be a sophisticated method, radiologists should be aware of all the pearls and pitfalls of DWI in specific diseases and conditions. That was the main concept when writing this book. We aimed to focus on the gastrointestinal applications of DWI in this book, in order to provide a comprehensive coverage of all technical and clinical aspects of MR imaging of the gastrointestinal system. We are fortunate to have had the support and contribution of an outstanding group of internationally renowned authors in this field. Their expertise, cooperation, and effort, which have made this book possible, are greatly appreciated.

Athens, Greece   Sofia Gourtsoyianni
Lisbon, Portugal   Nikolaos Papanikolaou

Contents

DWI Techniques and Methods for GI Tract Imaging

Thierry Metens and Nickolas Papanikolaou

1.1 Introduction

Diffusion is a mass transport process which results in molecular or particle mixing without requiring bulk motion [1]. For human tissues, water mobility can be assessed in the intracellular, extracellular and intravascular spaces. All media have a different degree of structure and thus pose a variant level of difficulty in water mobility that is called "diffusivity". An MRI pulse sequence sensitized to microscopic water mobility using strong gradient pulses can be utilized to provide insights into the complexity of the environment which in turn can reveal information related to tissue microarchitecture.

A major requirement in diffusion imaging is to select ultrafast pulse sequences that may freeze macroscopic motion in the form of respiration, peristalsis or patient motion in general. For this reason, echo-planar imaging (EPI) sequences modified with the addition of two identical strong diffusion gradients are routinely used to provide diffusion information. The amplitude and duration of the diffusion gradients are represented by the "b value" (measured in s/mm^2), an index used to control the sensitivity of DWI contrast to water mobility.

T. Metens
Department of Radiology, HôpitalErasme, MRI Clinics, Bruxelles, Belgium
e-mail: thierry.metens@erasme.ulb.ac.be

N. Papanikolaou (✉)
Oncologic Imaging, Computational Clinical Imaging Group Champalimaud Foundation, Centre for the Unknown, Lisbon, Portugal
e-mail: npapan@npapan.com

1.2   Basic Principles

1.2.1   Formal Definition of Diffusion

Molecules are involved in a random thermal motion called the Brownian motion, according to the observation in 1827 by Robert Brown [1773–1858] of the erratic translation movement of pollen in water, a random movement with no net ensemble displacement. This situation was further studied in 1908 by Paul Langevin [1872–1946] [2] and by Albert Einstein [1879–1955] [3]. This microscopic movement is due to thermal agitation and occurs for water molecules in a bath of pure water (self-diffusion) or a viscous liquid medium.

For the free three-dimensional diffusion, considering the Brownian motion under thermal agitation, Albert Einstein derived in 1905 the relation between the mean quadratic displacement, the diffusion coefficient and the diffusion time t:

$$\left\langle r^2 \right\rangle = 6Dt \tag{1.1}$$

In other words, starting from a position r_0 after a time t, the particles reach a standard deviation position located on the surface of a sphere of radius $(6Dt)^{1/2}$. The value of the diffusion coefficient D depends on the temperature T and the friction F (that is proportional to the viscosity) of the medium.

In living tissues, diffusion is restricted by many other factors like intracellular metabolites, the presence of cell membranes, the extracellular architecture, the relative size of cells and extracellular compartment. Therefore, the measured diffusion coefficient is called apparent diffusion coefficient (ADC). The apparent diffusion coefficient value is in general reduced if cells expand because of cytotoxic oedema or when the cell density is more elevated, like in most malignant tissues. The link between the ADC and tissue cellularity seems quite complex and is still under investigation [4].

In some highly organized tissues, anatomically wise water diffusion can be spatially restricted by the presence of ordered structures. Therefore, diffusion becomes anisotropic where the mathematical description requires a diffusion tensor D to be introduced. However, in the gastrointestinal tract, the majority of studies deal with the isotropic part of the diffusion tensor, i.e. the average diffusion measured in three orthogonal directions, called the average diffusivity or the mean diffusion. In what follows we shall simply refer to it as the diffusion coefficient.

1.2.2   Pulse Sequence Considerations

Following the seminal works on MR and diffusion by H. Carr and E. Purcell (1954) [5], H. Torrey (1956) [6] and D. Woessner (1961) [7], in 1965, E. Stejskal and J. Tanner [8] showed that the MR signal can be made sensitive to diffusion by the addition of supplementary gradients, called diffusion gradients (Fig. 1.1). Diffusing spins (moving spins) travelling at least partially along the direction of the diffusion gradients will accumulate a net dephasing, and this results into a signal attenuation, while stationary spins will be identically dephased and rephased with no signal loss. The Stejskal-Tanner gradients are used within a spin-echo echo-planar imaging (SE-EPI) sequence, allowing to acquire diffusion-weighted images (DWI). The signal of the SE Stejskal-Tanner sequence can be calculated as

$$S(TE) = S(TE = 0, b = 0)e^{-TE/T2}e^{-bD} \qquad (1.2)$$

with the diffusion control parameter b (in s/mm^2):

$$b = (\gamma G \delta)^2 (\Delta - \delta / 3) \qquad (1.3)$$

γ being the proton gyromagnetic ratio, G the gradient amplitude, Δ the delay between successive diffusion gradients and δ the duration of the diffusion gradients. The signal of the Stejskal-Tanner sequence is thus both T2 weighted and diffusion weighted; the b factor controls the diffusion weighting, and TE controls the T2 weighting (Fig. 1.2). The product of the two exponential attenuations explains the inherent low signal-to-noise ratio (SNR) of DWI; the spatial resolution is generally kept low in order to compensate for the otherwise low SNR.

Equation (1.2) accounts for diffusion in one specific direction in space and shows that increased water mobility results in substantial signal attenuation on diffusion-weighted images. Conversely, water molecules with reduced mobility will present with significantly lower signal attenuation compared to water molecules with increased mobility leading to a relative higher signal on high b value images. We emphasize that the signal intensity in DW images is not only affected by the b factor and the diffusion of water but also by the T2 and T2* relaxation time of the tissues because the Stejskal-Tanner diffusion "carrying sequence" is an SE-EPI. In a tissue with a long T2 relaxation coefficient, a relatively high signal intensity can be maintained mimicking restricted diffusion patterns, the so-called "T2 shine-through" effect. On the contrary, a tissue with a very low T2 value will appear dark, the so-called T2 shading effect. It is important to avoid such confusion by comparing DW images with T2-weighted images.

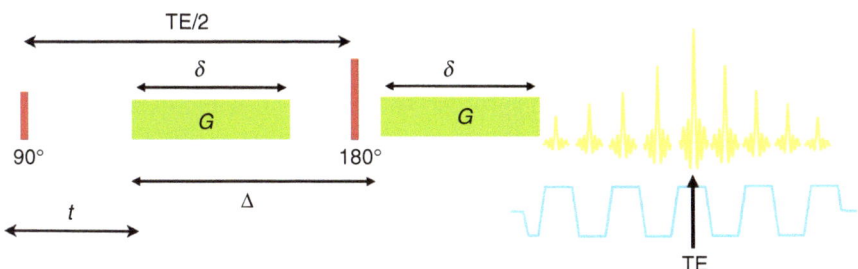

Fig. 1.1 Stejskal-Tanner SE diffusion sequence with EPI reading (only the diffusion-sensitized gradients are shown in green; these gradients are aligned along one spatial direction. G is the gradient amplitude, Δ is the delay between successive diffusion gradients and δ is the duration of the diffusion gradients; the 90° and 180° RF pulses are used to generate a spin echo in order to minimize T2* effects. Note that after the 180 RF pulse, the effective gradient sign is changed

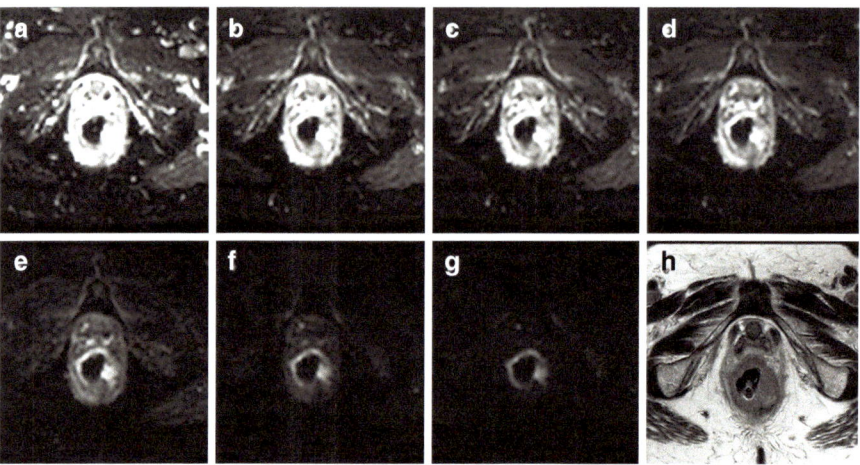

Fig. 1.2 (**a–h**) DW images acquired at fixed TE with $b = 0, 50, 100, 200, 500, 1000$ and 1500 s/mm^2 showing various degrees of diffusion weighting in a patient with rectal cancer. Note that the tumoural areas are preserving high signal intensities on the high b value images. Axial T2w TSE image on the same level with diffusion is shown as a reference

In many clinical situations, visual interpretation of DW images is not enough and further quantification of ADC is considered mandatory. This calculation involves at least the acquisition of two signals S1 and S2 from acquisitions with different b factors, i.e. b_1 and b_2 factors, and by computing

$$ADC = Ln\left(S1/S2\right)/\left(b_2 - b_1\right) \tag{1.4}$$

If more than two b value images are involved, a linear regression of the logarithms $Ln[S(b)/S(0)]$ in function of the b values will provide the ADC value (i.e. slope of the regression line). When this is performed for each pixel, a calculated image of the ADC, called the ADC map, is reconstructed. Generally the ADC map can be reconstructed by considering one or a combination of several diffusion directions and several b values, while the correct choice of the regression points is influencing the final ADC value. The calculated numerical ADC value depends on many parameters, and therefore we emphasize the "apparent" denomination.

The diffusion sequence must be repeated using diffusion gradients oriented in at least three orthogonal directions. The geometric mean of three orthogonal diffusion-weighted images with the same b amplitude gives the isotropic diffusion image (where directional effects have been eliminated by definition):

$$I\left(b\right) = S\left(0\right)e^{-(b_{xx}D_{xx} + b_{yy}D_{yy} + b_{zz}D_{zz})/3} = S\left(0\right)e^{-b(D_{xx} + D_{yy} + D_{zz})/3} = S\left(0\right)e^{-bMD} \tag{1.5}$$

The derivation of Eq. (1.2) is based on the hypothesis that the diffusion is the single source of intravoxel incoherent motion (IVIM). However, in highly perfused tissues, micro-perfusion represents another potential source of IVIM: the blood flow appears indeed random as it follows the randomly oriented capillaries, and during the diffusion time, spins in the capillary blood flow might have changed their direction several times, or different spins will flow along different directions in differently oriented capillaries. This micro-perfusion phenomenon constitutes a pseudo-diffusion movement and will be discussed in detail below. Another complication arises from the fact that the pixel size in DWI is large compared to the various tissue compartments with different diffusivity and partial volume effects might result, again justifying the apparent character of the diffusion coefficient measured in tissue.

The Stejskal-Tanner SE-EPI sequence is acquired using a single EPI echo train, providing an image in the so-called "single-shot" mode. In a segmented acquisition (multi-shot mode), the signal phase of the different k-space segments can interfere destructively causing an irremediable signal loss in the final image. This severe segmental dephasing occurs in the strong diffusion gradients because of several physiologic factors like heart pulsation and organ movement occurring in between consecutive k-space segments acquisitions. Segmented approaches are possible but require adequate phase corrections that are often difficult to be implemented. The SE-EPI sequence requires fat suppression to avoid the large water-fat shift in the EPI phase encoding direction. Moreover, T2* image blurring and spatial distortions due to the high EPI echo train length (ETL)-related artefacts are mostly present in the phase encoding direction (Fig. 1.3).

The spatial resolution of abdominal clinical DW imaging is in general relatively lower comparing to that of conventional MRI sequences like T1 or T2. However, multichannel high-density torso coils are nowadays available and allow the use of parallel imaging with high acceleration factors to minimize geometric distortions and T2* blurring (Fig. 1.4). High b value images suffer from poor SNR, and therefore more averages than in the low b values may be acquired to compensate for the limited signal. Depending on several factors including coil characteristics, phase encoding direction, size and shape of the object of interest, static magnetic field

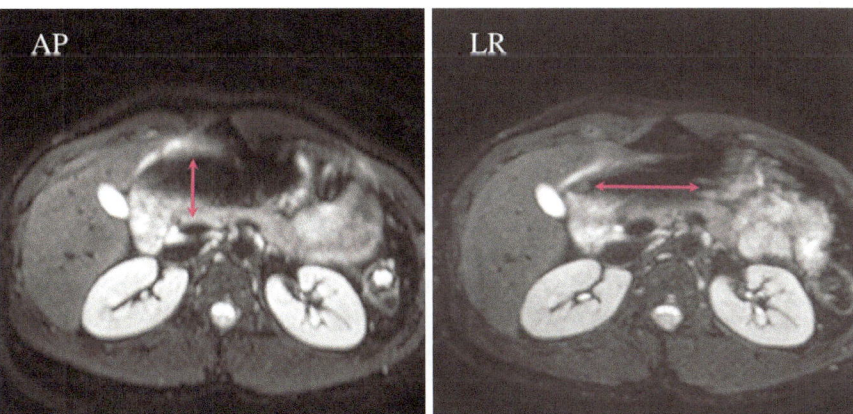

Fig. 1.3 DW images acquired at 3 T with b 10 s/mm^2 with phase encoding in the anterior to posterior direction (left) and in the left to right direction (right); the red arrows indicate the corresponding EPI spatial distortions that are pronounced on the stomach due to air presence

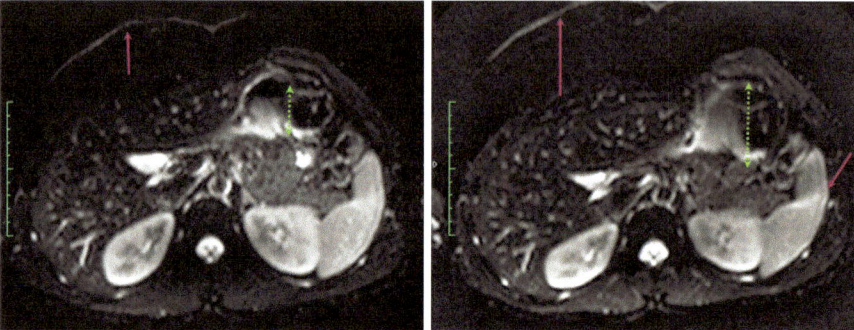

Fig. 1.4 SE-EPI images acquired at 3 T (same image matrix and TR, TE, with $b = 0$ s/mm^2) with anterior to posterior phase encoding direction, EPI echo train length of 41 (left) and 95 (right, with more blurring and larger distortions). The red arrows indicate the water-fat shift (distance between the skin and the liver, suboptimal fat suppression), and the green dotted arrow suggests the amplitude of the EPI spatial distortion. High parallel imaging acceleration factors enable lower echo train lengths that result in images with less artefacts

strength and image SNR, an acceleration factor between 2 and 4 is normally used. Recent progress has been provided by the multi-slice simultaneous excitation technique, which allows diffusion-weighted imaging of the liver accelerated by a factor 2 in addition to parallel imaging [9, 10]. However, its clinical utility hasn't been demonstrated on the gastrointestinal tract, yet.

1.2.3 Diffusion Modelling in GI Cancer

Malignant neoplasms often present with a strong perfusion component. The capillary random movement through the diffusion gradients constitutes an intravoxel incoherent motion, IVIM, dephasing the Stejskal-Tanner signal. An adequate model should thus take into account both water diffusion and the pseudo-diffusion from blood capillary flow. One approach is provided by the bi-exponential model [11–16]:

$$S(b)/S(0) = fe^{-bD*} + (1-f)e^{-bD} \qquad (1.6)$$

where f is the micro-perfusion fraction, $D*$ is the pseudo-diffusion coefficient linked to micro-perfusion and D is the water true diffusion coefficient in the tissue. The major disadvantage that prohibits routine clinical applications of IVIM is that the bi-exponential model is very prone to signal fitting errors. However, fortunately, in general $D* \gg D$, and there exists a value $b*$ that

$$\text{for } b > b^* : e^{-bD*} \simeq 0 \qquad (1.7)$$

meaning that for $b > b*$, the attenuation comes from pure diffusion D only. Thus, instead of trying at once to fit the bi-exponential model with three parameters f, $D*$ and D, an approximate strategy called "partial fitting" comprises D calculation by fitting a mono-exponential model to the signal of two or more images with b values larger than $b*$, typically $b* = 150$ s/mm^2:

$$\text{Ln}\left[S(b)/S(0)\right] \sim \text{Ln}(1-f) - bD = \beta - bD \tag{1.8}$$

where β is the intercept of the linear regression and

$$f = 1 - e^{\beta} \tag{1.9}$$

Alternatively, in a stepwise procedure, the previous approximate values obtained for b and D can be used as initial guessed values of the bi-exponential fit.

The kurtosis model is another non-mono-exponential decay of the diffusion-weighted signal [17–20]:

$$S(b)/S(0) = \exp-\left(bD + b^2 D^2 K/6\right) \tag{1.10}$$

Since diffusion is a process that takes place in four dimensions, we need to utilize a displacement distribution function that will predict the position of each water molecule at a certain time point. In a homogeneous medium, the molecular displacement distribution can be Gaussian, where the width is proportional to the diffusion coefficient. In heterogeneous tissues, the water molecular displacements significantly differ from the true "Brownian motion" defined for free molecules because water molecules bounce, cross, and interact with cell membranes and other microstructural components [17]. In the presence of these obstacles, the actual diffusion distance is reduced compared to free water, and the displacement distribution is no longer Gaussian (Fig. 1.5). In other words, while over very short times diffusion reflects the local intrinsic viscosity, at longer diffusion times, the effects of the obstacles become predominant.

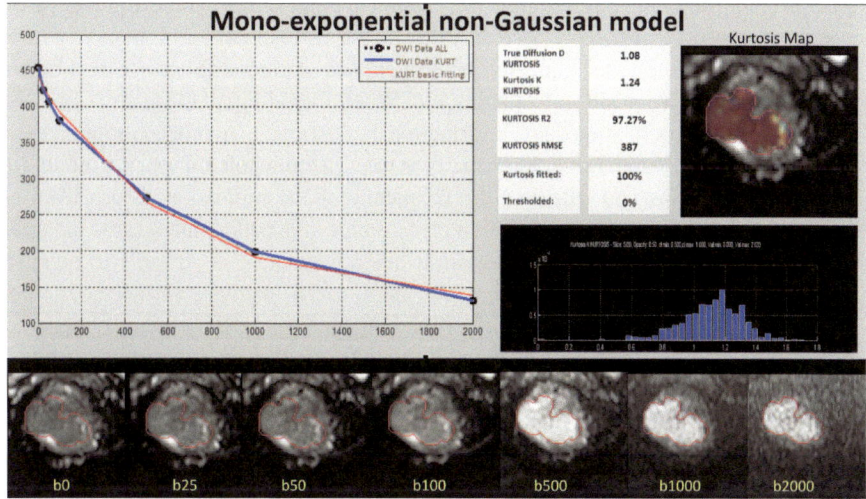

Fig. 1.5 Multi-b value DWI acquisition with up to a b value of 2000 s/mm² in a rectal tumour fitted with a non-Gaussian mono-exponential model taking into account tumour heterogeneity-related deviations on the very high b value area. The corresponding kurtosis map (K map) is showing the tumour to be highly kurtotic (mean $K = 1.24$), while the fitting accuracy of the corresponding model is very high (Adj $R^2 = 0.9727$)

1.2.4 Diffusion Biomarkers Quantification

The ADC value, most often derived from a mono-exponential model, depends on the signal fitting and the available SNR. Artefacts might also significantly influence the fit quality. The slope calculated by the linear regression, giving the ADC, will inevitably be influenced by the signal value measured for each point, especially the points corresponding to the highest b values because these have the most attenuated signal and lowest SNR, therefore are more prone to noise. The fit can be performed pixel by pixel to generate an ADC map but can also be based on ROI signal measurement and of course the definition of the ROI (size, position, homogeneous region or not) plays a crucial role [21]. Note that the mean ADC value in the ROI drawn on the ADC map is not identical to the ADC value computed from the mean signal of the same ROI drawn in all separate b value images because the Eq. (1.4) is non-linear. Globally we can expect a better fit quality if an increasing number of b values are used, at the cost of a longer acquisition duration. At fixed b value and diffusion direction, the acquisition might be repeated, providing an average from a higher number of samples. The number of b values or b directions might also be increased at acquisition; the choice of the b values involved in the fit will finally influence the ADC quantification (Fig. 1.6). A recent technique that can be used to further improve contrast between lesions and background tissue by further attenuating the signal of tissues that are bright on high b values is the so-called computed or synthetic b values (Fig. 1.7).

Various quantification strategies regarding the quantification of imaging biomarkers like ADC have been proposed. The traditional method is to draw a region of interest, either with irregular borders identifying the tumour or use multiple small circular regions of interest sampling many different areas of the tumour. The latter is recruited to overcome potential averaging effects when including the whole tumour in the region of interest from where a mean value of the corresponding biomarker will be calculated. Tumours are typically heterogeneous comprising necrosis, fibrosis, haemorrhage, active proliferative areas, etc. When including all these areas in a single region of interest, important information is destroyed. A more comprehensive way to tackle with tumoural heterogeneity is to perform whole tumour segmentation in three dimensions and calculate the frequency of the biomarker values from all pixels belonging to the tumour. Such a representation is called histogram, and apart from the possibility to visualize tumoural heterogeneity, it is possible to quantify various histogram metrics including min, max, mean, median, variance, standard deviation, percentiles, skewness, kurtosis and others (Fig. 1.8). By quantifying histogram metrics before and after treatment, someone can assess in a comprehensive way how the effect of treatment to the specific biomarker is utilized.

In the clinical setting, a compromise must be found considering the number of slices, the spatial resolution, the use of trigger methods or a breath-hold approach. Of course, a central question is whether the examination aims at lesion detection (ADC map with enough contrast between the lesion and the surrounding tissue) or aims at a precise quantification with low variability, required in the case of a longitudinal study meant to detect a change due to the natural evolution of the disease or being the signature of a response to a treatment.

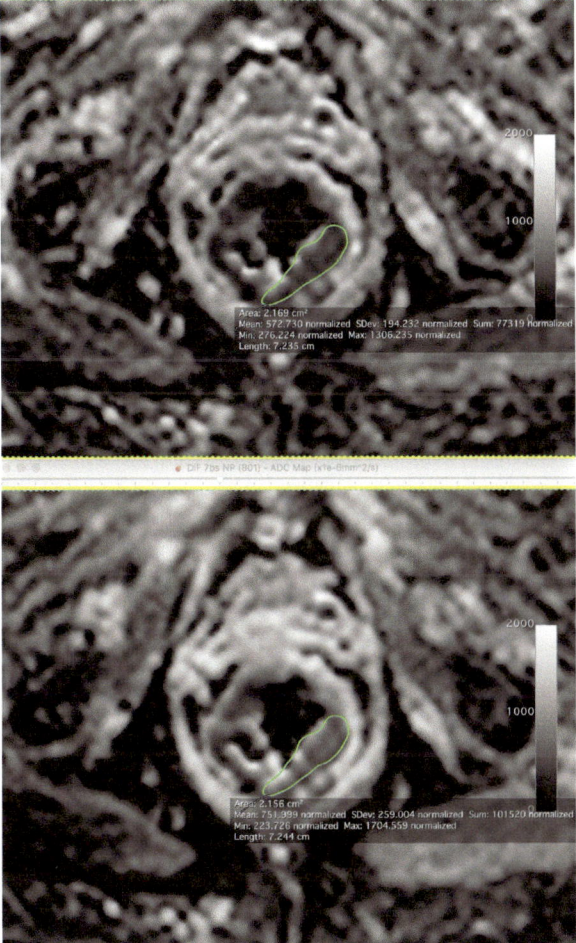

Fig. 1.6 ADC maps (units 10^{-6} mm^2/s) in a patient with rectal tumour obtained by signal fitting using $b = 200$, 500 and 1000 s/mm^2 (upper image) and by signal fitting using $b = 0$, 200, 500 and 1000 s/mm^2 (lower image). The tumoural ADC value measured on the lower image comparing to the higher image because it is influenced by the large micro-perfusion contamination that is evident when using low b value data in our mono-exponential fitting algorithms

1.3   Clinical Applications

According to ESGAR recommendations [22], use of diffusion-weighted imaging (DWI) in rectal cancer patients is not obligatory for primary baseline staging. DWI sequences, however, are implemented in routine rectal cancer MRI protocols nowadays. In tissues presenting with high cellular content and intact cell membranes, such as rectal tumour tissue, water motion is restricted. This does not apply to the tissue from which tumour originates, in our case rectal wall, and thus DWI is very

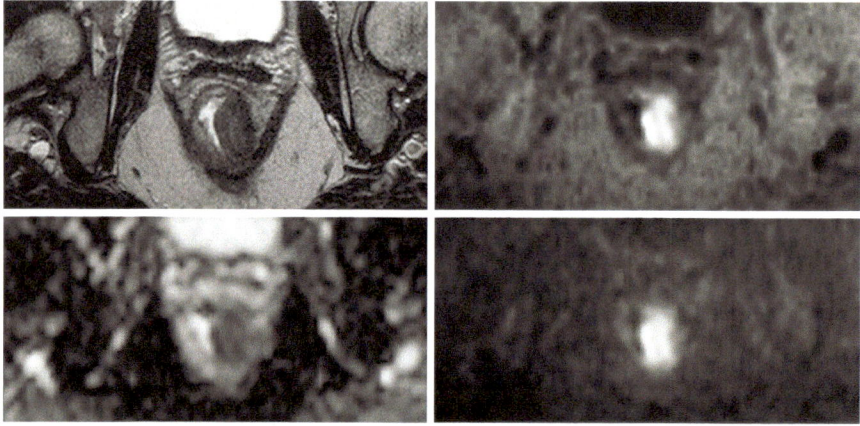

Fig. 1.7 Patient with rectal cancer, axial T2-weighted TSE (upper left), calculated b 1500 (upper right), ADC (lower left) and originally acquired b 1500 (lower right). Calculated b value image successfully demonstrates the high signal intensity of the tumour with better SNR as compared to originally acquired b 1500 image

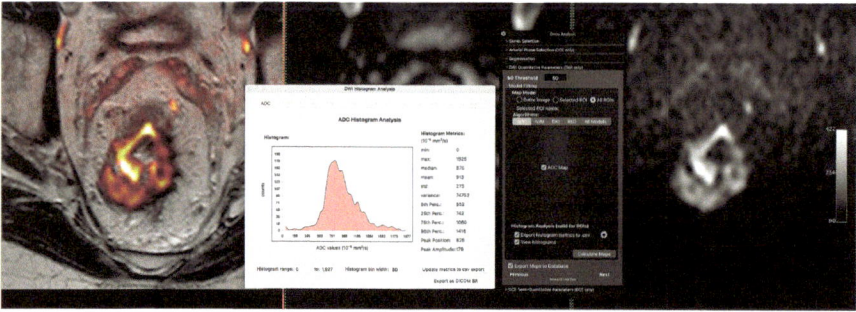

Fig. 1.8 Whole tumour histogram analysis in a patient with rectal cancer. Histogram metrics are shown together with the tumour histogram and can be used to assess treatment-related changes (i.e. chemoradiation effects)

helpful in tumour detection. DWI sequences for clinical purposes are relatively quick to perform, do not require the administration of contrast medium, and can be integrated to the existing rectal tumour imaging protocol without significant delay in the overall examination time. Optimally, DWI images have to be obtained with the same slice thickness and orientation as the high-resolution axial oblique T2-weighted images. For quantitative analysis or measurements for assessment of therapy response, at least three b values should be obtained [23]. It is advised as there is no standardization regarding the use of DWI in the abdomen, and more specifically for rectal cancer imaging, each centre should test the apparent diffusion coefficient (ADC) values produced by their sequence or scanner and have own references.

Lack of DWI protocol standardization for post-CRT MRI leads to variability in ADC values measured, with no universal cut-off value for differentiating between viable tumour and complete response (CR). However, post-neoadjuvant treatment means ADC values of responding tumours have a tendency to increase. Using an axial-orientated diffusion-weighted sequence with background body signal suppression and b values, 0, 500, and 1000 s/mm^2, Curvo Semedo et al. [24] demonstrated that lower ADC values were associated with a more aggressive tumour profile. ADC has the potential to become an imaging biomarker of tumour aggressiveness profile. Kim et al. [25] reported that the diagnostic accuracy in evaluating CR significantly increases when DWI is combined with conventional MRI. However, restricted diffusion in the corresponding tumours has been reported in 42% of the patients who achieve pathological complete response (pCR) after neoadjuvant CRT, reasons for this according to surgical specimen pathology review being the presence of intramural mucin, degree of proctitis and mural fibrosis [26].

1.3.1 Whole-Body Diffusion

Whole-body MRI is already incorporated in the clinical routine due to recent technological advances in the field of RF technology including integrated high-density phased array coils that can cover large anatomical areas, providing an adequate signal-to-noise ratio. Integrated parallel imaging techniques accelerated image acquisition and resulted in significant reductions in examination times making whole-body MRI clinically feasible. The main area of clinical applications of whole-body MRI is screening for metastatic disease [27]. The workhorse sequence to detect metastatic disease is diffusion-weighted imaging due to the very high conspicuity that offers to detect hypercellular lesions. A major challenge in terms of image quality on diffusion-weighted images is efficient fat saturation that can be very difficult to achieve in large anatomic areas. The latter can be addressed to the presence of air in the gastrointestinal tract and the variable amount and distribution of fat due to different body habitus. Therefore, the mainstream pulse sequence that should be recruited for whole-body diffusion applications should provide robust fat suppression. Such a method should make use of inversion pulses to achieve signal nulling of the fat, the so-called STIR technique (short tau inversion recovery), which is well known to be less sensitive to Bo inhomogeneity. The diffusion sequence making use of inversion pulses to achieve fat saturation is called DWIBS (diffusion-weighted imaging with body suppression) and is usually acquired in axial plane with a single b value ranging between 600 and 1000 depending on the gradient performance of the MRI scanner.

Conclusions

Diffusion-weighted imaging of the gastrointestinal tract is technically challenging due to the presence of multiple physiologic motions (respiration, peristaltic motion), as well as the presence of air. With careful optimization of pulse sequence parameters, it has the potential to aid radiologists not only on differential diagnosis but also on assessing and predicting treatment response.

References

1. Basser O, Ozarslan E. Introduction to diffusion. In: Johansen-Berg H, Behrens TEJ, editors. Diffusion MRI: from quantitative measurement to in vivo neuroanatomy. Amsterdam: Elsevier; 2014.
2. Paul L. Sur la théorie du mouvement brownien. Comptes-Rendus de l'Académie des Sciences. 1908;146:530–2.
3. Albert E. Über die von der molekularkinetischen Theorie der Wärme geforderte Bewegung von in ruhenden Flüssigkeiten suspendierten Teilchen. Ann Phys. 1905;322(8):549–60.
4. White M, Dale A. Distinct effects of nuclear volume fraction and cell diameter on high b-value diffusion MRI contrast in tumours. Magn Reson Med. 2014;72:1435–43.
5. Carr HY, Purcell EM. Effects of diffusion on free precession in nuclear magnetic resonance experiments. Phys Rev. 1954;94:630–8.
6. Torrey HC. Bloch equations with diffusion terms. Phys Rev. 1956;104(3):563–5.
7. Woessner DE. Effects of diffusion in nuclear magnetic resonance spin-echo experiments. J Chem Phys. 1961;34:2057–61.
8. Stejskal EO, Tanner JE. Spin diffusion measurements: spin echoes in the presence of a time-dependent field gradient. J Chem Phys. 1965;42(1):288–92.
9. Gagoski SK, BA PJR, Witzel T, Wedeen VJ, Wald LL. Blipped-controlled aliasing in parallel imaging for simultaneous multislice echo planar imaging with reduced g-factor penalty. Magn Reson Med. 2012;67(5):1210–24.
10. Obele CC, Glielmi C, Ream J, Doshi A, Campbell N, Zhang HC, Babb J, Bhat H, Chandarana H. Simultaneous multislice accelerated free-breathing diffusion-weighted imaging of the liver at 3T. Abdom Imaging. 2015;40:2323–30.
11. Le Bihan D, Breton E, Lallemand D, Aubin ML, Vignaud J, Laval-Jeantet M. Separation of diffusion and perfusion in intravoxel incoherent motion MR imaging. Radiology. 1988;168:497–505.
12. Cohen AD, Schieke MC, Hohenwalter MD, Schmainda KM. The effect of low b-values on the intravoxel incoherent motion derived pseudodiffusion parameter in liver. Magn Reson Med. 2015;73:306–11.
13. Ahlgren A, Knutsson L, Wirestam R, Nilsson M, Ståhlberg F, Topgaard D, Lasič S. Quantification of microcirculatory parameters by joint analysis of flow-compensated and non-flow-compensated intravoxel incoherent motion (IVIM) data. NMR Biomed. 2016;29:640–9.
14. Lemke A, Stieltjes B, Schad LR, Laun FB. Toward an optimal distribution of b values for intravoxel incoherent motion imaging. Magn Reson Imaging. 2011;29(6):766–76.
15. Lemke A, Laun FB, Simon D, Stieltjes B, Schad LR. An in vivo verification of the intravoxel incoherent motion effect in diffusion-weighted imaging of the abdomen. Magn Reson Med. 2010;64(6):1580–5.
16. Wetscherek A, Stieltjes B, Laun FB. Flow-compensated intravoxel incoherent motion diffusion imaging. Magn Reson Med. 2015;74(2):410–9.
17. Jensen JH, Helpern JA, Ramani A, Lu H, Kaczynski K. Diffusional kurtosis imaging: the quantification of non-Gaussian water diffusion by means of magnetic resonance imaging. Magn Reson Med. 2005;53:1432–40.
18. Jansen JF, Stambuk HE, Koutcher JA, Shukla-Dave A. Non-Gaussian analysis of diffusion-weighted MR imaging in head and neck squamous cell carcinoma: a feasibility study. AJNR Am J Neuroradiol. 2010;31(4):741–8.
19. Wang F, Jin D, Hua XL, Zhao ZZ, Wu LM, Chen WB, Wu GY, Chen XX, Chen HG. Investigation of diffusion kurtosis imaging for discriminating tumors from inflammatory lesions after treatment for bladder cancer. J Magn Reson Imaging. 2017;
20. Dia AA, Hori M, Onishi H, Sakane M, Ota T, Tsuboyama T, Tatsumi M, Okuaki T, Tomiyama N. Application of non-Gaussian water diffusional kurtosis imaging in the assessment of uterine tumors: a preliminary study. PLoS One. 2017;12(11):e0188434.

21. Colagrande S, Pasquinelli F, Mazzoni LN, Belli G, Virgili G. MR-diffusion weighted imaging of healthy liver parenchyma: repeatability and reproducibility of apparent diffusion coefficient measurement. J Magn Reson Imaging. 2010;31:912–20.
22. Beets-Tan RGH, Lambregts DMJ, Maas M, Bipat S, Barbaro B, Curvo-Semedo L, Fenlon HM, Gollub MJ, Gourtsoyianni S, Halligan S, Hoeffel C, Kim SH, Laghi A, Maier A, Rafaelsen SR, Stoker J, Taylor SA, Torkzad MR, Blomqvist L. Magnetic resonance imaging for clinical management of rectal cancer: updated recommendations from the 2016 European Society of Gastrointestinal and Abdominal Radiology (ESGAR) consensus meeting. Eur Radiol. 2018;28:1465. https://doi.org/10.1007/s00330-017-5026-2.
23. Le Bihan D, Poupon C, Amadon A, Lethimonnier F. Artifacts and pitfalls in diffusion MRI. J Magn Reson Imaging. 2006;24(3):478–88.
24. Curvo-Semedo L, Lambregts DM, Maas M, Beets GL, Caseiro-Alves F, Beets-Tan RG. Diffusion-weighted MRI in rectal cancer: apparent diffusion coefficient as a potential noninvasive marker of tumor aggressiveness. J Magn Reson Imaging. 2012;35(6):1365–71. https://doi.org/10.1002/jmri.23589.
25. Kim SH, Lee JM, Hong SH, Kim GH, Lee JY, Han JK, Choi BI. Locally advanced rectal cancer: added value of diffusion-weighted MR imaging in the evaluation of tumor response to neoadjuvant chemo- and radiation therapy. Radiology. 2009;253(1):116–25.
26. Kim SH, Lee JY, Lee JM, Han JK, Choi BI. Apparent diffusion coefficient for evaluating tumour response to neoadjuvant chemoradiation therapy for locally advanced rectal cancer. Eur Radiol. 2011;21(5):987–95.
27. Lecouvet FE, Mouedden El J, Collette L, et al. Can whole-body magnetic resonance imaging with diffusion-weighted imaging replace Tc 99m bone scanning and computed tomography for single-step detection of metastases in patients with high-risk prostate cancer? Eur Urol. 2012;62:68–75.

Upper Gastrointestinal Tract

2

Mirna Al-Khouri, Adel Abdellaoui, and Simon Jackson

2.1 Introduction

Whilst many patients with suspected upper gastrointestinal tract pathology undergo initial endoscopic assessment, multimodality imaging techniques which include contrast imaging studies, ultrasound, multi-detector computed tomography (MDCT) and positron emission tomography-computed tomography (PET-CT) also play an important role in patient management particularly for the diagnosis and staging of patients with suspected malignancy. Traditionally MRI has not offered significant advantages when compared to the other imaging modalities used in this anatomical area due to a range of technical limitations. However, recent technological developments have resulted in new clinical applications which include the potential role of quantitative DWI as a biomarker for the diagnosis and assessment of upper GI tract malignancy including treatment response. In this chapter, we will cover the evolving role of DWI in the upper GI tract including relevant technical details and clinical applications as well as possible future directions in patient care.

2.2 Technical Details

2.2.1 Patient Preparation/Protocols

The recommendation to fast patients for at least 4–6 h prior to the MRI examination is widely accepted. Prior to the examination, a MRI safety checklist should be completed for each patient. Patients are routinely positioned in the supine position with

M. Al-Khouri · A. Abdellaoui · S. Jackson (✉)
Department of Radiology, Derriford Hospital, Plymouth Hospitals NHS Trust, Plymouth, Devon, UK
e-mail: malkhouri@nhs.net; adel.abdellaoui@nhs.net; simon.jackson1@nhs.net

© Springer International Publishing AG, part of Springer Nature 2019
S. Gourtsoyianni, N. Papanikolaou (eds.), *Diffusion Weighted Imaging of the Gastrointestinal Tract*, https://doi.org/10.1007/978-3-319-92819-7_2

the feet pointing towards the magnet, although the supine position with head-first position is also practised [1].

Body coils provide a larger cover of surface area and better homogeneity. However, they have poorer signal-to-noise ratio (SNR) as well as contrast-to-noise ratio (CNR) when compared to surface coils which are widely used in the abdominal imaging. Smaller external coils such as phased-array cardiac coils have been used and repositioned depending on the site of tumour, to ensure the tumour is located within the middle distance of the coil [2]. Cardiac and respiratory triggering for imaging of the upper GI tract can be useful in minimising motion artefact. When ECG triggering is used, the R wave is used to trigger an RF pulse.

Many patient preparation protocols are used to achieve sufficient gastric distension. Most studies describe using 500–1000 mL of water for adequate luminal distension. Despite some variability, attempts should be made to maintain the reproducibility of visceral distension.

Giganti et al. used water with Ferumoxsil as a negative oral contrast agent prior to imaging. A number of naturally occurring negative oral agents, such as blueberry and pineapple juice, are being readily used in MR imaging of the abdomen and pelvis with superior results in comparison to those without an agent [3]. The results are presumed to be due to the paramagnetic effect of the relatively high concentration of manganese, especially with pineapple juice.

Published studies have used antispasmodic medication to minimise the effect of gastric peristalsis [1, 4]. In the absence of contraindications, intramuscular Buscopan (hyoscine butylbromide, Boehringer Ingelheim Ltd) is routinely administered.

2.2.2 Image Acquisition

The magnet field strength used in most centres is 1.5 T, although studies are performed in 3 T MRI scanners [5, 6]. The standard examination protocol for upper GI MR imaging includes both T2-weighted sequences and DWI. Images are traditionally acquired in the axial plane. Coronal views can be used to assess the extent of the tumour. T1-weighted fat-suppressed images with and without contrast are also obtained.

Diffusion-weighted images in the upper GI tract can be obtained using a free-breathing, multi-averaging technique or a single-shot, breath-hold technique. The latter is more frequently used as it provides a rapid assessment of the target organ, with good anatomic detail and less susceptibility to motion artefact [7]. The slice thickness can vary from 3 to 5 mm and reach up to 10 mm to cover the entire length of the oesophagus.

The more recent development of diffusion-weighted whole-body imaging with background body signal suppression (DWIBS) provides the feasibility for acquiring DWI with free breathing, and although recent advances allow the use of functional diffusion analysis in DWIBS to produce an ADC map, the accuracy and reproducibility of the technique are still under investigation [8].

The routine b-values used in upper gastrointestinal imaging range between 0 and 1000 s/mm^2. Many centres use a b-value of 0 with additional intermediate and high diffusion gradients including values of 500 and 800 s/mm^2 or 600 and 1000 s/mm^2 for the oesophagus, stomach and periampullary region. From our experience, and in agreement with published research, we found that a *b*-value of 800 s/mm^2 provides a good balance between diffusion and image contrast.

Wang et al. demonstrated that most accurate tumour length measurements of the oesophagus were obtained using $b = 600$ s/mm^2 [9]. Whereas Lee et al. found that in the periampullary region, lesions are more conspicuous at *b*-value of 800 s/mm^2 demonstrating higher signal intensity, as bile appeared more frequently hyperintense on $b = 500$ s/mm^2 than $b = 800$ s/mm^2 [10]. The clinical application and interpretation of DWI in the upper gastrointestinal tract remain to be standardised to minimise disparity in the quantitative findings among different institutions.

2.3   Artefact and Image Optimization

As discussed previously, MRI of the upper gastrointestinal tract is considered technically challenging due to motion artefact from the cardiac, respiratory and unpredictable physiological peristalsis. Flow artefact arising from adjacent aortic and pulmonary vasculature can diminish the accuracy of image interpretation. Movement artefact from the heart and lungs can be diminished by using automatic gated navigators.

Furthermore, due to the central location of the mediastinum in the body, the sensitivity of the receiver coil is reduced, and the signal-to-noise (SNR) ratio is subsequently diminished. An increase in the magnet strength (3 T in comparison to 1.5 T) will improve the SNR; however, a higher degree of susceptibility artefact will be encountered due to the larger surface area of air-fluid-tissue interface in the thoracic cavity [11].

Patients undergoing MRI to evaluate the upper GI tract may have an oesophageal stent in situ. Most oesophageal stents are 'MRI compatible' at 1.5 T; however, some are indicated conditional for use either due to MRI-related heating or generation of artefact or both. Image acquisition with a controlled whole-body averaged specific absorption rate (SAR) is generally recommended for conditional stents, although the manufacturer's reference manual for magnetic resonance safety recommendations should be referred to for each implant or device on individual cases (Table 2.1).

2.4   Clinical Applications

2.4.1   Upper GI Tract Malignancy

2.4.1.1 The Oesophagus

Oesophageal cancer remains a leading cause of cancer-related mortality, being the sixth most common cause of death worldwide accounting for 400,000 deaths in 2012 [12, 13]. The two most common histological subtypes are squamous cell carcinoma and adenocarcinoma. Whilst adenocarcinoma has become increasingly

Table 2.1 Staging with pretreatment ADC

Study	Location of tumour	Patient preparation prior to imaging	ADC cut-off	b-values (s/mm^{-2})—magnet strength	Main outcome	Additional comments
Zhang et al. [4]	$n = 23$ BT4 gastric cancer, $n = 23$ healthy volunteers	Overnight fasting, 800–1000 mL water	$ADC < 1.84 \times 10^{-3}$ mm²/s was a cut-off value to distinguish BT4 gastric cancer from poorly distended stomach wall (specificity 95.7%, sensitivity 100%)	0, 1000–1.5 T	Mean ADC value for BT4 gastric cancer was $1.12 \pm 0.23 \times 10^{-3}$ mm²/s; significantly lower than nearby normal stomach wall of $2.11 \pm 0.21 \times 10^{-3}$ mm²/s ($p < 0.001$)	Mean ADC value of the normal stomach in healthy volunteers was $1.93 \pm 0.22 \times 10^{-3}$ mm²/s, which was significantly higher than that of BT4 gastric cancer ($p < 0.01$)
Giganti et al. [15]	Oesophageal or Siewert I $n = 18$ (nine surgery and nine CRT before imaging)	500 mL of water and Ferumoxsil	Optimal cut-off for local invasion: mean $ADC = 1.33 \times 10^{-3}$ mm²/s ($p = 0.05$)	0, 600–1.5 T	MR has high specificity (92%) and accuracy (83%) for T staging and high sensitivity (100%) with moderate accuracy (66%) for N staging	ADC values were different between surgery-only and chemo-/radiotherapy groups (1.90 mm²/s vs. 1.30×10^3 mm²/s, respectively; $p = 0.005$)
Aoyagi et al. [17]	Oesophageal SCC $n = 123$ ($n = 31$ surgery, $n = 84$ neoadjuvant therapy, $n = 8$ not treated)	6 h clear fluid only. N-Butyl scopolamine bromide IM	1.5×10^{-3} mm²/s (92% sensitivity, 86% specificity, 89% accuracy)	0, 1000–1.5 T	ADC values of cancer were significantly lower than normal oesophagus $(1.145 \pm 0.321 \times 10^{-3}$ mm²/s vs. $2.001 \pm 0.385 \times 10^{-3}$ mm²/s, respectively; $p < 0.0001$)	ADC values of advanced-stage tumours were significantly lower than early-stage tumours, and a significant difference between Stages I and III ($p < 0.05$), I and IV ($p < 0.05$), II and III ($p < 0.05$) and II and IV ($p < 0.05$) was observed

Study	Population	Preparation	ADC value	b-values / field strength	Findings	Conclusions
Hou et al. [20]	Oesophageal SCC n = 42 (all had radical surgery)	12-h fasting. Supine	–	400, 600, 800–1.5 T	Oesophageal SCC lengths are most precise on DWI when compared with CT or MRI. Fused DWI/CT images were used to improve accuracy to delineate gross tumour volume (GTV)	Oesophageal SCC GTV upper and lower margins were clearly depicted; therefore, DWI fused with CT images can be used for radiation treatment planning systems
Avcu et al. [23]	Gastric cancer n = 70 and healthy individuals n = 30	-	1.12×10^{-3} mm²/s to differentiate malignant from benign gastric wall thickening (100% sensitivity, 98.6% specificity)	50, 400, 800–1.5 T	The difference between ADC values of adenocarcinoma and lymphoma was statistically significant, $0.85 \pm 0.16 \times 10^{-3}$ mm²/s and $1.09 \pm 0.08 \times 10^{-3}$ mm²/s, respectively ($p < 0.05$)	No statistical significance was found between the subtypes of gastric carcinoma
Kantarci et al. [24]	Gastric cancer n = 21	-	0.982 mm²/s (87% sensitivity, 100% specificity)	50, 400, 800–1.5 T	Mean ADC values of gastric tumours were significantly lower than normal gastric wall, 0.892 ± 0.23 SD mm²/s and 1.453 ± 0.35 SD mm²/s, respectively	Mean ADC values of gastric tumours were significantly lower than those found in the normal gastric wall ($p < 0.05$)
Lee et al. [28]	Biliary strictures in the periampullary region n = 78	–	–	0, 500, 800–1.5 T	DWI can help differentiate between malignant and benign periampullary lesions. Most periampullary carcinomas appear hyperintense on high b value DWI	$b = 800$ s/mm² is the optimal b value for periampullary regions. DWI added to MRCP improves the diagnostic accuracy for ampullary lesions

prevalent in Western countries, squamous cell carcinoma demonstrates the highest incidence worldwide [13].

Accurate initial tumour staging is mandatory for determining optimal patient management. A number of classification systems have been developed with the UICC TNM classification offering internationally agreed standards to categorise tumours and provide an indication for patient prognosis. The most recent version of the *TNM Classification of Malignant Tumours* (8th edition) for oesophageal cancer is shown in Table 2.2 [14].

Multimodality imaging plays an important role in the accurate staging of oesophageal tumours with MDCT, PET-CT and endoscopic ultrasound (EUS) widely recommended in accepted guidelines. MRI offers an alternative noninvasive cross-sectional modality which can accurately delineate anatomical detail of the oesophagus. A recent study by Giganti et al. [15] showed promising results in which MRI with added DWI had the highest specificity (92%) and accuracy (83%) for T staging when compared with other modalities [15]. In addition, DWI MR had the highest sensitivity (100%) in the detection of nodal disease with moderate accuracy (66%).

The optimal ADC cut-off values between normal oesophagus and cancer tissue can vary between studies, due to the lack of a standardised protocol. ADC cut-off values between 1.3 and 1.5×10^{-3} mm^2/s have been described [15–17]. In the study by Aoyagi et al. [16], there was a good correlation between the ADC value and the clinical T and N stages in patients with oesophageal SCC tumours with the ADC value much lower in more advanced disease. In addition, there was also a negative correlation between ADC value, tumour diameter and SUV [16].

Angiogenesis remains an important element in determining tumour growth and risk of metastatic spread. Previous studies have found a negative correlation of ADC

Table 2.2 TNM staging of oesophageal cancer

Stage	Level of involvement
T—primary tumour	
TX	Primary tumour cannot be assessed
T0	No evidence of primary tumour
Tis	Carcinoma in situ/high-grade dysplasia
T1a	Tumour invades lamina propria or muscularis mucosae
T1b	Tumour invades submucosa
T2	Tumour invades muscularis propria
T3	Tumour invades adventitia
T4a	Tumour invades pleura, pericardium, azygous vein, diaphragm or peritoneum
T4b	Tumour invades other structures such as aorta, vertebrae or trachea
N—Regional lymph nodes	
NX	Regional lymph nodes cannot be assessed
N0	No regional lymph node metastasis
N1	Metastasis in one to two regional lymph nodes
N2	Metastasis in three to six regional lymph nodes
N3	Metastasis in seven or more regional lymph nodes
M—Distant metastasis	
M1	No distant metastasis
M0	Distant metastasis

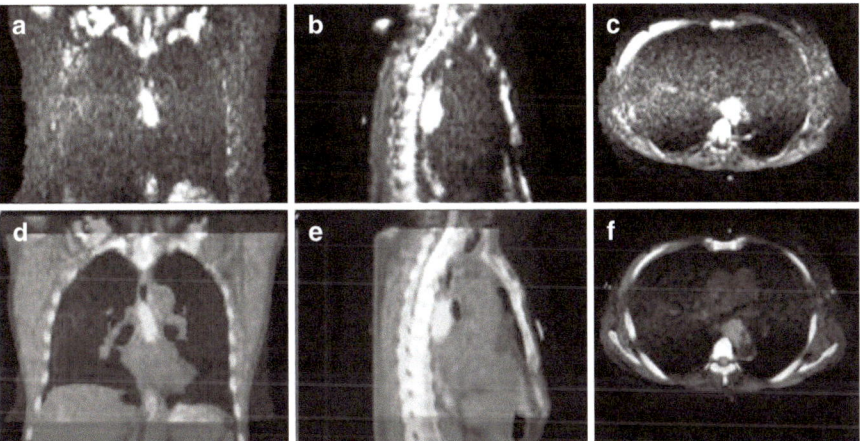

Fig. 2.1 DWI scans and fused images. (**a**), (**b**), and (**c**) show the coronal, sagittal and transverse images of DWI scans; (**d**), (**e**), and (**f**) show the coronal, sagittal and transverse images of fused images. Hou et al. [20]

values with vascular endothelial growth factor (VEGF) which itself is a biomarker for tumour angiogenesis and prognosis. This explains the correlation between low ADC values and more advanced disease [17].

Tumour length is also an independent predictor of survival in patients with oesophageal cancer [18]. A study published by Hou et al. demonstrated that diffusion imaging is more precise than CT and conventional MRI in delineating oesophageal SCC length and calculating gross tumour volume. The difference between the pathology specimen was −0.54 ± 6.03 mm at b-value of 600 when compared with CT (3.63 ± 12.06 mm) and T2-weighted images (3.46 ± 11.41 mm). In addition, DWI images can also be fused with CT for radiation treatment planning to allow more focused therapy (Fig. 2.1) [19].

2.4.1.2 The Stomach

Gastric cancer is a leading cause of death worldwide accounting for 754,000 deaths in 2015 [20]. Multimodality imaging techniques including ultrasound, CT, MRI and PET-CT, as well as endoscopic assessment, are also widely performed to detect, stage and follow-up patients with gastric malignancy. The most recent version of the *TNM Classification of Malignant Tumours* (8th edition) for staging gastric cancer is shown in Table 2.3 [14].

A number of studies have shown that DWI and ADC can provide useful information in differentiating benign and malignant gastric pathology, thus offering an important role for MRI, especially in patients who refuse or cannot tolerate endoscopy [21]. A range of ADC cut-off values between 0.982 and 1.12×10^{-3} mm^2/s have been described [22–24]. In similarity with oesophageal malignancy, the range can be explained by the absence of standardised protocols and variations between magnet strength and vendors (Figs. 2.2, 2.3, 2.4, 2.5, 2.6, 2.7, and 2.8).

Table 2.3 TNM staging of gastric cancer

Stage	Level of involvement
T—primary tumour	
TX	Primary tumour cannot be assessed
T0	No evidence of primary tumour
Tis	Carcinoma in situ/high-grade dysplasia
T1a	Tumour invades lamina propria or muscularis mucosae
T1b	Tumour invades submucosa
T2	Tumour invades muscularis propria
T3	Tumour invades subserosa
T4a	Tumour perforates serosa
T4b	Tumour invades other structures
N—Regional lymph nodes	
NX	Regional lymph nodes cannot be assessed
N0	No regional lymph node metastasis
N1	Metastasis in one to two regional lymph nodes
N2	Metastasis in three to six regional lymph nodes
N3	Metastasis in seven or more regional lymph nodes
M—Distant metastasis	
M1	No distant metastasis
M0	Distant metastasis

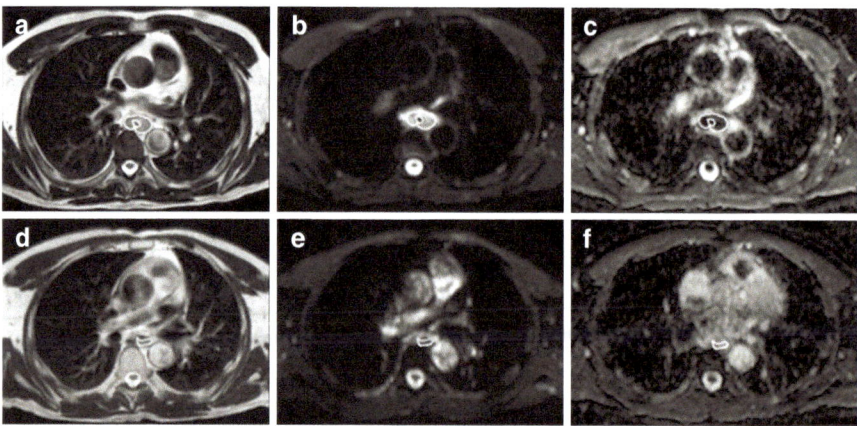

Fig. 2.2 Axial T2 image showing a lesion of the middle oesophageal wall. (**a**) ROI measured on DWI (*b* value, 600 mm/s^2) (**b**) along the lesion border and then copied into the ADC map (**c**) calculating a pre-NT ADC value of $1.21 \pm 0.26 \times 10^{-3}$ mm^2/s. Restaged after NT, a reduction was observed in the wall thickening (**d**) drawing an ROI (*b* value, 600 mm/s^2) (**e**) and then copying it to the map; a significant rise in the ADC values was observed ($2.11 \pm 0.33 \times 10^{-3}$ mm^2/s) (**f**). This patient was found to be a responder with TRG 2. Courtesy of De Cobelli et al. Eur Radiology (2013) [31]

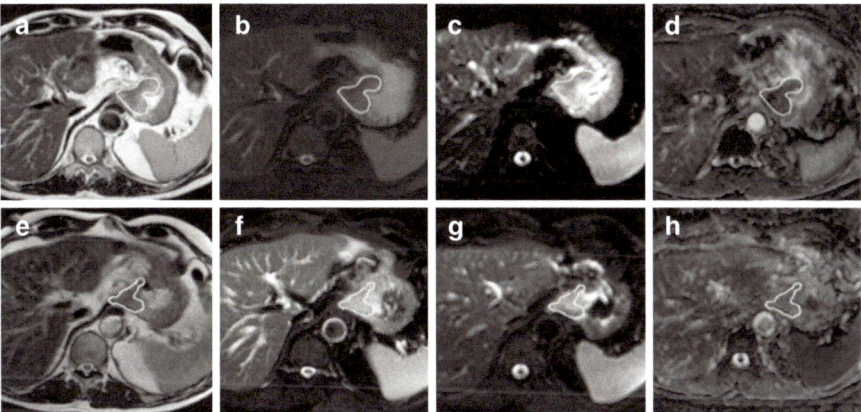

Fig. 2.3 Axial T2 with (**a**) and without (**b**) fat suppression showing a lesion of the subcardial region extended to the gastric fundus; an ROI was drawn on DWI (*b*-value, 600 mm/s^2) along the lesion border (**c**) and then copied to the ADC map (**d**) calculating a pre-NT ADC value of $1.73 \pm 0.29 \times 10^{-3}$ mm^2/s. Restaged after NT, a slight reduction of the wall thickening was observed (**e, f**); drawing an ROI (*b*-value of 600 mm/s^2) (**g**) and copying to the map, no significant rise in the ADC values was observed ($1.58 \pm 0.23 \times 10^{-3}$ mm^2/s) (**h**). This patient was found to be a nonresponder with TRG 4. Courtesy of De Cobelli et al. Eur Radiology (2013) [31]

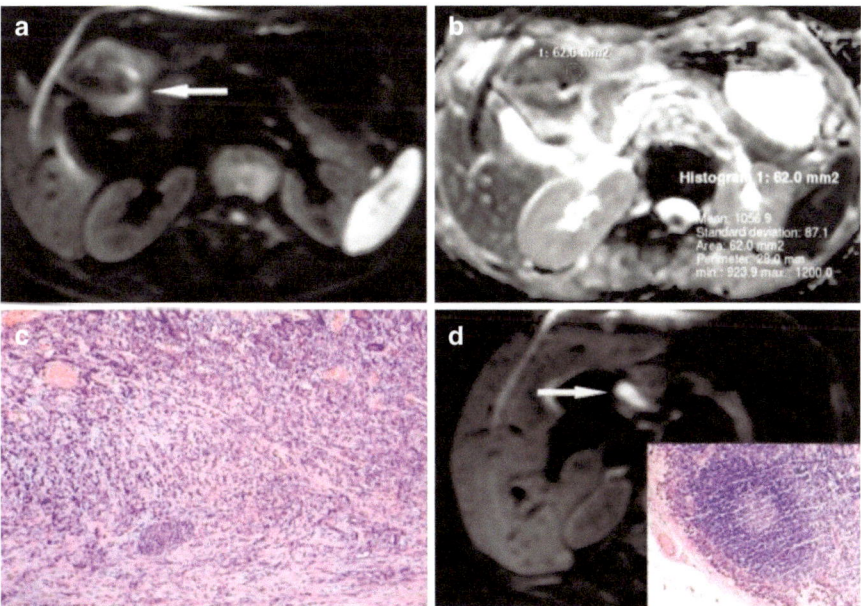

Fig. 2.4 Female, 69 years old, with gastric carcinoma pathologically diagnosed at stage T3N3aM0. (**a**) Axial DWI (*b* = 1000 s/mm^2) shows the lesion with remarkably high signal intensity at the antrum (arrow) of the stomach. (**b**) Corresponding ADC map shows restricted mean and minimum ADC values of the lesion as 1056.9 and 923.9 × 10^{-6} mm^2/s, respectively. (**c**) The photomicrograph (haematoxylin and eosin staining, × 100) shows a poorly differentiated adenocarcinoma with a Lauren classification of diffuse type. (**d**) Axial DWI of this patient in the upper slice shows a bright lymph node (arrow) metastasis proved microscopically (embedded picture, haematoxylin and eosin staining; original magnification, ×100) [1]. Images used with permission of Wiley & Sons, Inc.

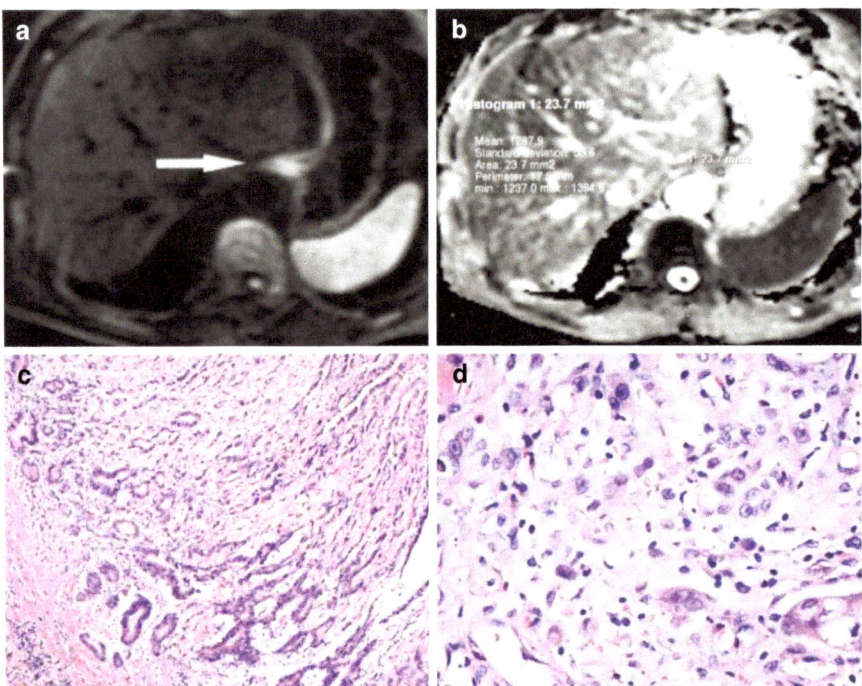

Fig. 2.5 Male, 61 years old, with gastric carcinoma pathologically diagnosed at stage T2N0M0. (a) Axial DWI (b = 1000 s/mm^2) shows the lesion with remarkably high signal intensity at the cardia (arrow) of the stomach. (b) Corresponding ADC map shows restricted mean and minimum ADC values of the lesion as 1287.9 and 1237.0 $\times$ 10^{-6} mm^2/s, respectively. (c) The photomicrograph (haematoxylin and eosin staining; original magnification, $\times$100) shows a poor moderately differentiated adenocarcinoma with a Lauren classification of mixed type. (d) The photomicrograph (haematoxylin and eosin staining; original magnification, $\times$400) of the lesion shows the cellular atypia such as large cell column, increased nucleus/cytoplasm ratio and irregular cell shape [1]. Images used with permission of Wiley & Sons, Inc.

In one study 3 T DWI-MRI shows an overall accuracy of 82.9% for the T staging of gastric cancer. This reaches 100% in T4 disease, whilst the N staging accuracy is 74.3% [6].

There is a significant difference between the ADC values of the histological subgroups of adenocarcinomas reflecting the degree of cellularity, with lower ADC values noted relating to the poorly differentiated type in comparison to the moderately and well-differentiated types [6].

MRI with DWI has a diagnostic accuracy similar to that of conventional MRI and MDCT. However, for the N staging, DWI-MRI has a higher sensitivity but lower specificity (86.7 and 58.8, respectively) [5]. Furthermore, DWI is superior to T2-weighted sequence in the detection of early T1 gastric cancer (91.6% vs. 75%) and staging of more advanced cancers (accuracy of 87.9% vs. 69.7%) [23]. There is also a significant difference in the ADC values of adenocarcinoma and lymphoma [23].

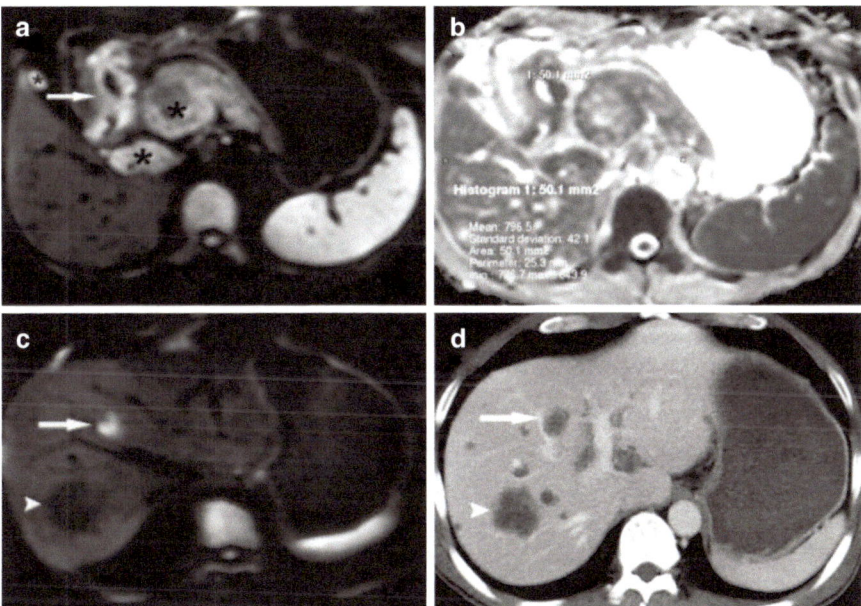

Fig. 2.6 Male, 50 years old, with gastric carcinoma pathologically diagnosed at stage T4N1M1. (**a**) Axial DWI ($b = 1000$ s/mm^2) shows the lesion with remarkably high signal intensity at the antrum (arrow) of the stomach with multiple enlarged metastatic lymph nodes (asterisks). (**b**) Corresponding ADC map shows restricted mean and minimum ADC values of the lesion as 796.5 and 726.7×10^{-6} mm^2/s, respectively. (**c**) Axial DWI of the same patient in the upper slice shows a liver metastasis with high signal intensity (arrow) proved by needle biopsy. Note the liver cyst posterior with low signal intensity (arrowhead). (**d**) Axial contrast enhancement CT image in the venous phase of the same patient shows both hepatic lesions with equivocal rim enhancement, which could not be differentiated from each other [1]. Images used with permission of Wiley & Sons, Inc.

In one study ADC has shown to be a strong independent prognostic indicator in evaluating patients with gastric cancer. An ADC value of 1.5×10^{-3} mm^2/s or lower has been associated with a poor prognosis in all patients with gastric cancer in both chemotherapy and surgery groups [2].

Tomizawa et al. found that DWIBS/T2 has a similar sensitivity to PET-CT in the diagnosis of upper gastrointestinal cancers. Their patient group included gastric, oesophageal, GIST and duodenal cancers, but most cases were gastric [24].

2.4.1.3 Ampullary and Periampullary Tumours

Ampullary tumours are usually small and difficult to identify on cross-sectional imaging. Jang et al. compared 39 patients with benign ampullary obstruction and 23 patients with biopsy-proven malignant ampullary tumours. They found that adding DWI can significantly increase the diagnostic accuracy of conventional 3 T MRI with high sensitivity and specificity. The accuracy improved from 85–87% to 97–98% with a sensitivity of 91–96% and specificity of 100%. The mean ADC value (1.23×10^{-3} mm^2/s) was also significantly lower in the malignant group [25].

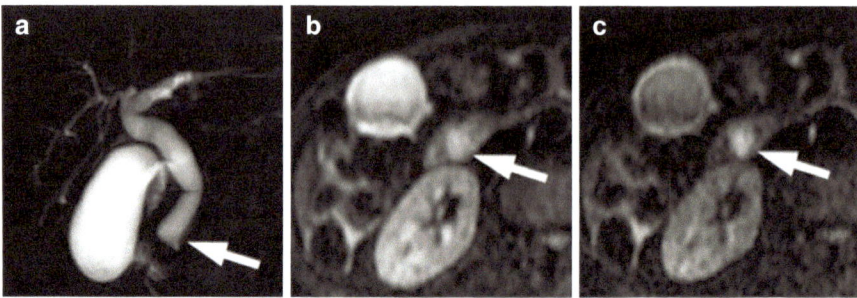

Fig. 2.7 Periampullary carcinoma in a 69-year-old man. (**a**) MRCP demonstrating biliary duct dilatation (arrow). $b = 500$ s/mm^2 DWI (**b**) and $b = 800$ s/mm^2 (**c**) [28]. Springer publication

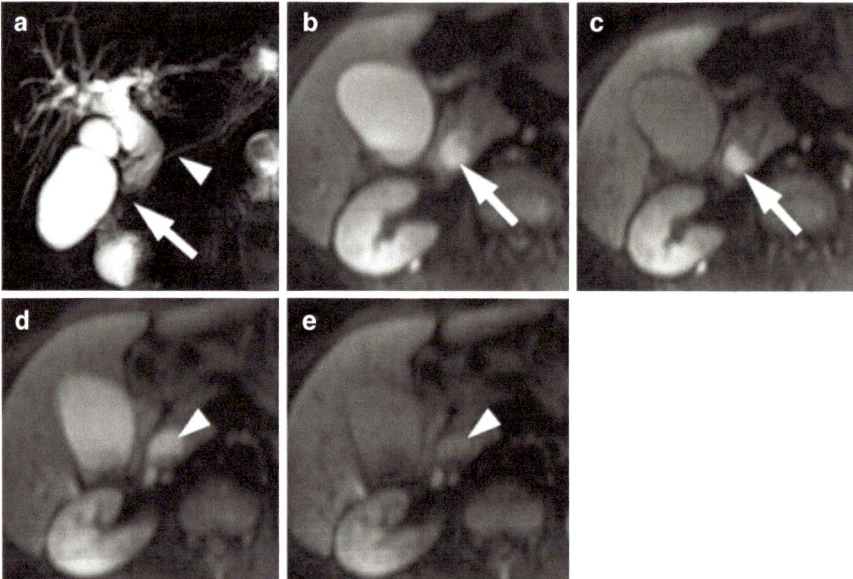

Fig. 2.8 Ampullary carcinoma in a 63-year-old man. (**a**) MRCP shows marked bile duct dilatation with a mass at the ampullary region (arrow), and main pancreatic duct dilatation (arrowhead) was noted. $b = 500$ s/mm^2 (**b**) and $b = 800$ s/mm^2 (**c**) show a hyperintense ampullary mass (arrows). (**d**) $b = 500$ s/mm^2 DWI superior to (**b**) shows a hyperintense bile (arrowhead). (**e**) $b = 800$ s/mm^2 DWI superior to (**c**) shows decreased signal intensity of bile (arrowhead) [28]. Springer publication

Lee et al. found that combining DWI with MRCP improved the diagnostic accuracy for the differentiation between benign and malignant periampullary strictures. The accuracy improved from 0.71 to 0.92. A b-value of 800 s/mm^2 was suggested as the optimal value for demonstrating restricted diffusion on DWI [10].

Diagnostic confidence is also further improved by viewing standard anatomic sequences with diffusion-weighted images [26].

2.4.2 Role of DWI in Treatment Response

Predicting the response to neoadjuvant therapy (nCRT) at an early stage reduces the exposure of patients to ineffective treatment. DWI and ADC interpretation has been proven useful in predicting treatment response and survival in patients with oesophageal SCC [27–29]. On DWI, the disappearance of high signal or hyperintensity expression at 1–3 months post-treatment is also associated with increased overall survival in this group of patients [30]. Therapies which target tumour vasculature result in reduced ADC values, especially when interrogated using low b-values, which are sensitive to vascular perfusion effects [28–30]. This is seen particularly in the initial 24 h of treatment, due to the increase in the intracellular water and loss of the extracellular space which result in a transient decrease in ADC value [33].

Patients with gastro-oesophageal cancers receiving neoadjuvant therapy were found to have significantly lower pretreatment ADC values. This implies that ADC values can predict pathological response and presurgery neoadjuvant therapy can be justified for this group of patients, to downstage tumour size prior to surgery. One exploratory study did not find an association between pretreatment ADC and predicting treatment response in patients with oesophageal cancer; however, the treatment-induced change in ADC (ΔADC) during the first 2–3 weeks of nCRT was highly predictive of the histopathological response [29]. The discrepancy of results found in the literature can be explained by the significant difference between the ADC values of adenocarcinomas and SCC, the level of tumoural differentiation and the use of differing scanning protocols between institutes. Table 2.4 illustrates the methodology and results of studies investigating treatment response.

Similar findings were observed in patients with advanced T4 oesophageal SCC, where ΔADC after radiotherapy of 15% predicted responders with an accuracy of 85% [30].

It has been concluded that ADC changes are more reliable than dimensional criteria in assessing oesophageal tumours, and as such ADC assessment can optimise management of locally advanced gastro-oesophageal cancers [31].

2.4.3 Other Upper GI Pathologies

2.4.3.1 Gastrointestinal Lymphoma

Over the past decade, the applications of DWI have been increasingly studied in oncological settings, particularly for the detection of lymphoma [32]. Due to normal lymph node anatomy, reactive lymph nodes can restrict diffusion and therefore may display variable degrees of signal intensity on diffusion-weighted imaging. The cut-off point in the literature has not been established between normal, reactive and malignant nodes.

The stomach is the most common site for extra-nodal lymphoma, whereas primary oesophageal lymphoma accounts for <1% of all GI lymphomas [33].

Table 2.4 Response monitoring ADC

Study	Number of patients and location of tumour	Treatment	Patient preparation prior to imaging	ADC cut-off	b-values (s/mm^{-2})—Magnet strength	Main outcome	Comments
Giganti et al. [2]	28 GEC[a] and 71 gastric cancers	Surgery-alone $n = 71$ and nCRT[b]s $n = 28$ pretreatment ADC values were acquired	500 mL of water and Ferumoxsil antispasmodic and contrast administered following patient positioning	Mean ADC value 1.5×10^{-3} mm^2/s or lower were associated with a negative prognosis ($p = 0.002$)	0, 600–1.5 T	ADC could represent a noninvasive quantitative parameter that is potentially helpful in evaluating the aggressiveness of gastric cancer	Lower ADC values were associated with the use of neoadjuvant chemotherapy
Aoyagi et al. [32]	80 patients with oesophageal SCC	Pre-nCRT ADC values were acquired	-	Mean ADC value for oesophageal cancers was $1.10 \pm 0.28 \times 10^{-3}$ mm^2/s (range 0.36–1.86) ADC cut-off point 1.10×10^{-3} mm^2/s	0, 1000–1.5 T	A high ADC was associated with better response to CRT than low ADC ($p < 0.01$)	A low ADC value was an independent risk factor for lower survival rate ($p = 0.04$)
Rossum et al. [37]	20 patients with oesophageal cancer	Pre-, during and post-nCRT ADC values were acquired	Supine position, no antiperistaltic agents administered		0, 200, 800–1.5 T	There is a significant association between ΔADC_{during} and pathological response	A low ΔADC during of <21% predicted a poor pathologic response (specificity and PPV 100%)
De Cobelli et al. [39]	31 locally advanced GEC	Pre- and post-neoadjuvant treatment	300–500 mL water for visceral distension and Ferumoxsil, scopolamine butylbromide IM. Contrast	Post-treatment ADC cut-off: 1.84×10^{-3} mm^2/s to differentiate responders from nonresponders	0, 600–1.5 T	Post-ADC values may help to discriminate responders and nonresponders	Patients with post-NT ADC values above the cut-off are responders (sensitivity = 70.6%, specificity = 80%, $p = 0.0007$)

[a]GEC: gastro-oesophageal cancer including junctional cancers
[b]nCRT: neoadjuvant chemoradiotherapy

Diffusion-weighted imaging has been proven to be useful in the detection of gastric lymphoma, which displays restriction of diffusion on high b-value and a low signal intensity on the calculated ADC map. Furthermore, in differentiating between gastric lymphoma and adenocarcinoma, the mean ADC value was found to be statistically significant by Avcu et al. with lower ADC values associated with cancer [23].

2.4.3.2 Stromal Tumours

Gastrointestinal stromal tumours (GISTs) are mesenchymal tumours of the gastrointestinal tract. The stomach is the most common site reported with approximately 60% of GISTs being gastric in origin, although they may arise from anywhere along the GI tract [34]. GISTs have a malignant potential and can still recur after excision.

Accurate risk stratification is important for the selection of patients who would benefit from adjuvant treatment. A study by Kang et al. demonstrated that an ADC cut-off value of 1.279×10^{-3} mm^2/s could be used as a biomarker to differentiate the grade of GISTs, with 100% sensitivity, a moderate specificity of 62.2% and an overall accuracy of 81.8%. However, tumour size and necrosis did not show a significant difference [42].

2.4.3.3 Inflammation

Crohn's disease rarely affects just the stomach and isolated gastric involvement accounts for <0.07% [35]. Patients with typical presentation of inflammatory bowel disease undergo serological investigations, endoscopy and histological testing to confirm the diagnosis. MR small bowel/enterography with the addition of DWI is well established and will be covered in the small bowel chapter.

Conclusion

Diffusion-weighted MR imaging is a noninvasive modality which can provide functional and quantitative assessment of tissues, without the burden of radiation and intervention or the need for extracorporeal contrast agents. High signal intensity on DWI with relatively low quantitative values on the ADC map was found to be able to distinguish benign from malignant upper GI tract disease, with a reliable diagnostic accuracy as discussed in this chapter. Generally, tumours with higher ADC values are thought to be associated with a better prognosis.

The role of diffusion imaging in the upper gastrointestinal tract is still expanding, and its future direction in patient care is promising; however, standardisation of protocols and further advances in technology are required to increase the clinical confidence and reliability of DWI in both oncological and non-oncological applications.

References

1. Liu S, Wang H, Guan W, Pan L, Zhou Z, Yu H, et al. Preoperative apparent diffusion coefficient value of gastric cancer by diffusion-weighted imaging: correlations with postoperative TNM staging. J Magn Reson Imaging. 2015;42(3):837–43.
2. Giganti F, Orsenigo E, Esposito A, Chiari D, Salerno A, Ambrosi A, et al. Prognostic role of diffusion-weighted MR imaging for resectable gastric cancer. Radiology. 2015;276(2):444–52.
3. Ali AF. Comparative study of pineapple juice as a negative oral contrast agent in magnetic resonance cholangiopancreatography. J Clin Diagn Res. 2015;9(1):13–6.
4. Zhang X, Tang L, Sun Y, Li Z, Ji J, Li X, et al. Sandwich sign of Borrmann type 4 gastric cancer on diffusion-weighted magnetic resonance imaging. Eur J Radiol. 2012;81(10):2481–6.
5. Joo I, Lee JM, Kim JH, Shin CI, Han JK, Choi BI. Prospective comparison of 3T MRI with diffusion-weighted imaging and MDCT for the preoperative TNM staging of gastric cancer. J Magn Reson Imaging. 2015;41(3):814–21.
6. Liang J, Lv H, Liu Q, Li H, Wang J, Cui E. Role of diffusion-weighted magnetic resonance imaging and apparent diffusion coefficient values in the detection of gastric carcinoma. Int J Clin Exp Med. 2015;8(9):15639.
7. Sinha R, Rajiah P, Ramachandran I, Sanders S, Murphy PD. Diffusion-weighted MR imaging of the gastrointestinal tract: technique, indications, and imaging findings. Radiographics. 2013;33(3):655–76.
8. Kwee TC, Takahara T, Ochiai R, Nievelstein RA, Luijten PR. Diffusion-weighted whole-body imaging with background body signal suppression (DWIBS): features and potential applications in oncology. Eur Radiol. 2008;18(9):1937–52.
9. Wang L, Liu L, Han C, Liu S, Tian H, Li Z, Ren X, Shi G, Wang Q, Wang G. The diffusion-weighted magnetic resonance imaging (DWI) predicts the early response of esophageal squamous cell carcinoma to concurrent chemoradiotherapy. Radiother Oncol. 2016;121(2):246–51.
10. Lee NK, Kim S, Seo HI, Kim DU, Woo HY, Kim TU. Diffusion-weighted MR imaging for the differentiation of malignant from benign strictures in the periampullary region. Eur Radiol. 2013;23(5):1288.
11. van Rossum PS, van Lier AL, Lips IM, Meijer GJ, Reerink O, van Vulpen M, Lam MG, van Hillegersberg R, Ruurda JP. Imaging of oesophageal cancer with FDG-PET/CT and MRI. Clin Radiol. 2015;70(1):81–95.
12. Hong SJ, Kim TJ, Nam KB, Lee IS, Yang HC, Cho S, Kim K, Jheon S, Lee KW. New TNM staging system for esophageal cancer: what chest radiologists need to know. Radiographics. 2014;34(6):1722–40.
13. Thrift AP. The epidemic of oesophageal carcinoma: where are we now? Cancer Epidemiol. 2016;41:88–95.
14. Brierley JD. TNM classification of malignant tumours. New York, NY: John Wiley & Sons; 2017.
15. Giganti F, Ambrosi A, Petrone MC, Canevari C, Chiari D, Salerno A, Arcidiacono PG, Nicoletti R, Albarello L, Mazza E, Gallivanone F. Prospective comparison of MR with diffusion-weighted imaging, endoscopic ultrasound, MDCT and positron emission tomography-CT in the pre-operative staging of oesophageal cancer: results from a pilot study. Br J Radiol. 2016;89(1068):20160087.
16. Aoyagi T, Shuto K, Okazumi S, Shimada H, Nabeya Y, Kazama T, Matsubara H. Evaluation of the clinical staging of esophageal cancer by using diffusion-weighted imaging. Exp Ther Med. 2010;1(5):847–51.
17. Aoyagi T, Shuto K, Okazumi S, Hayano K, Satoh A, Saitoh H, Shimada H, Nabeya Y, Kazama T, Matsubara H. Apparent diffusion coefficient correlation with oesophageal tumour stroma and angiogenesis. Eur Radiol. 2012;22(6):1172–7.
18. Eloubeidi MA, Desmond R, Arguedas MR, Reed CE, Wilcox CM. Prognostic factors for the survival of patients with esophageal carcinoma in the US. Cancer. 2002;95(7):1434–43.
19. Hou D-L, Shi G-F, Gao X-S, Asaumi J, Li X-Y, Liu H, Yao C, Chang JY. Improved longitudinal length accuracy of gross tumor volume delineation with diffusion weighted magnetic resonance imaging for esophageal squamous cell carcinoma. Radiat Oncol. 2013;8(1):169.

20. Cancer [Internet]. World Health Organization. 2017 [cited 5 March 2017]. Available from: http://www.who.int/mediacentre/factsheets/fs297/en/.
21. Onur MR, Ozturk F, Aygun C, Poyraz AK, Ogur E. Role of the apparent diffusion coefficient in the differential diagnosis of gastric wall thickening. J Magn Reson Imaging. 2012;36(3):672–7.
22. Avcu S, Arslan H, Unal O, Kotan C, Izmirli M. The role of diffusion-weighted MR imaging and ADC values in the diagnosis of gastric tumors. J Belgian Soc Radiol. 2012;95(1)
23. Liu S, He J, Guan W, Li Q, Zhang X, Mao H, Yu H, Zhou Z. Preoperative T staging of gastric cancer: comparison of diffusion-and T2-weighted magnetic resonance imaging. J Comput Assist Tomogr. 2014;38(4):544–50.
24. Tomizawa M, Shinozaki F, Uchida Y, Uchiyama K, Fugo K, Sunaoshi T, Ozaki A, Sugiyama E, Baba A, Fukamizu Y, Kagayama S. Diffusion-weighted whole-body imaging with background body signal suppression/T2 image fusion and positron emission tomography/computed tomography of upper gastrointestinal cancers. Abdom Radiol. 2015;40(8):3012.
25. Jang KM, Kim SH, Lee SJ, Park HJ, Choi D, Hwang J. Added value of diffusion-weighted MR imaging in the diagnosis of ampullary carcinoma. Radiology. 2013;266(2):491–501.
26. Ichikawa T, Erturk SM, Motosugi U, Sou H, Iino H, Araki T, Fujii H. High-b value diffusion-weighted MRI for detecting pancreatic adenocarcinoma: preliminary results. Am J Roentgenol. 2007;188(2):409–14.
27. Wang L, Han C, Zhu S, Shi G, Wang Q, Tian H, Kong J, Zhang A. Investigation of using diffusion-weighted magnetic resonance imaging to evaluate the therapeutic effect of esophageal carcinoma treatment. Oncol Res Treat. 2014;37(3):112–6.
28. Morani AC, Elsayes KM, Liu PS, Weadock WJ, Szklaruk J, Dillman JR, Khan A, Chenevert TL, Hussain HK. Abdominal applications of diffusion-weighted magnetic resonance imaging: where do we stand. World J Radiol. 2013;5(3):68.
29. van Rossum PS, van Lier AL, van Vulpen M, Reerink O, Lagendijk JJ, Lin SH, van Hillegersberg R, Ruurda JP, Meijer GJ, Lips IM. Diffusion-weighted magnetic resonance imaging for the prediction of pathologic response to neoadjuvant chemoradiotherapy in esophageal cancer. Radiother Oncol. 2015;115(2):163–70.
30. Imanishi S, Shuto K, Aoyagi T, Kono T, Saito H, Matsubara H. Diffusion-weighted magnetic resonance imaging for predicting and detecting the early response to chemoradiotherapy of advanced esophageal squamous cell carcinoma. Dig Surg. 2013;30(3):240–8.
31. De Cobelli F, Giganti F, Orsenigo E, Cellina M, Esposito A, Agostini G, Albarello L, Mazza E, Ambrosi A, Socci C, Staudacher C. Apparent diffusion coefficient modifications in assessing gastro-oesophageal cancer response to neoadjuvant treatment: comparison with tumour regression grade at histology. Eur Radiol. 2013;23(8):2165.
32. Lin C, Luciani A, Itti E, Haioun C, Safar V, Meignan M, Rahmouni A. Whole-body diffusion magnetic resonance imaging in the assessment of lymphoma. Cancer Imaging. 2012;12(2):403.
33. Lo Re G, Federica V, Midiri F, Picone D, La Tona G, Galia M, Lo Casto A, Lagalla R, Midiri M. Radiological features of gastrointestinal lymphoma. Gastroenterol Res Pract. 2015;2016:2498143.
34. Kang TW, Kim SH, Jang KM, Choi D, Ha SY, Kim KM, Kang WK, Kim MJ. Gastrointestinal stromal tumours: correlation of modified NIH risk stratification with diffusion-weighted MR imaging as an imaging biomarker. Eur J Radiol. 2015;84(1):33–40.
35. Ingle SB, Hinge CR, Dakhure S, Bhosale SS. Isolated gastric Crohn's disease. World J Clin Case. 2013;1(2):71.

Small Bowel

3

Sonja Kinner

3.1 Introduction

Unlike the upper and lower gastrointestinal tract, the diagnostic access to the small bowel is not as easy. Therefore, it is of utmost importance to find a diagnostic tool to evaluate this more "hidden" gastrointestinal part with high sensitivity and specificity. Indeed, capsule endoscopy is now available for diagnosing the small bowel, but this tool is cost-intensive and there are certain contraindications, especially bowel strictures or obstruction [1]. Furthermore, capsule endoscopy only allows visualization of the surface, which is the mucosal layer, and deeper tissues cannot be evaluated, not to mention the surrounding tissues and other abdominal organs. Over the past years, cross-sectional imaging techniques like computed tomography (CT) and magnetic resonance imaging (MRI) have evolved considerably and changed the diagnostic approach of the small bowel. Both techniques are now optimized for small bowel imaging and play a more and more increasing role in the diagnosis of small bowel pathologies. CT is burdened with a still not negligible amount of radiation exposure, which is especially of importance for patients who are to undergo multiple examinations during their life due to the recurrent character of their underlying disease. As a result, MR imaging has become an increasingly used diagnostic tool and nowadays is an important imaging modality for evaluating small bowel pathologies. Besides the already in-place sequences like balanced steady state free precession, T2-weighted images with and without fat saturation and T1-weighted images unenhanced and dynamically acquired after the injection of a gadolinium-based contrast agent, diffusion-weighted imaging (DWI) has gained more and more importance in the last decade within the MR imaging protocol of the small bowel [2]. As DWI relies on the diffusion of water in tissue and

S. Kinner
Department of Diagnostic and Interventional Radiology and Neuroradiology,
University Hospital Essen, Essen, Germany
e-mail: sonja.kinner@uk-essen.de

© Springer International Publishing AG, part of Springer Nature 2019
S. Gourtsoyianni, N. Papanikolaou (eds.), *Diffusion Weighted Imaging of the Gastrointestinal Tract*, https://doi.org/10.1007/978-3-319-92819-7_3

therefore on tissue microarchitecture, it provides functional quantitative information and helps to distinguish diseased and normal tissue and therefore to assess therapy response [3]. DWI is a non-contrast technique, which is of extreme importance when it comes to imaging patients in whom the injection of a gadolinium-based contrast agent should be omitted. This is the case in pregnant patients, patients with renal impairment and also at times for children. In general, with the recent discussion of gadolinium deposition in the brain [4], each contrast agent injection should be carefully considered, and if DWI proves to be able to replace contrast-enhanced imaging, it should be used as a first-line technique.

3.2   Prerequisites

3.2.1   Patient Preparation

For small bowel MR imaging, patients should fast for 4–6 h at least. If only the small bowel is of interest, a thorough cleansing protocol does not have to be followed. Generally, there are two ways to achieve distension of the small bowel. Patients can drink a hyperosmolar solution about 45–60 min prior to the examination (MR enterography). For the alternate technique, MR enteroclysis, the solution is given via a nasoenteric tube, which is placed under fluoroscopic guidance before MR imaging [5]. This technique requires for the patient to change location, is more time-consuming and is burdened with associated tube placement radiation exposure. Some studies show that MR enteroclysis is able to distend the small bowel loops a little better than MR enterography [6–8]; others did not find a difference [9].

In our institution, we perform MR enterography and use an in-house pharmacy-made solution of locust bean gum and mannitol (LBG-mannitol). The patient is asked to drink at least 1 L of the solution before imaging [10]. Other oral solutions (containing polyethylene glycol, barium sulphate or alike) have been tested and perform similarly [11].

Recent studies suggest that DWI can be performed without bowel preparation reliably if looking at the large bowel [12]. For small bowel imaging, we still believe that bowel distension is of extreme importance as collapsed bowel loops mostly show higher signal in DWI and can therefore mimic pathologies, which have also been discussed by Li et al. in their paper. In their opinion poor loop distension results in apparent thickening and signal increase of the collapsed bowel segment mimicking inflammation on standard sequences and DWI [13].

To suppress peristalsis and concomitant artefacts, intravascular or intramuscular agents have to be injected. We use hyoscine butylbromide, also known as scopolamine butylbromide (Buscopan, Boehringer Ingelheim, Ingelheim, Germany), which is injected intravenously and has a half-life of 5 h. If contraindications for scopolamine are present (glaucoma, benign prostate hyperplasia) or in countries where butylscopolamine is not approved, glucagon (GlucaGen, Novo Nordisk Inc., Princeton, New Jersey) can be used [14]. The half-life is much shorter with 8–18 min but sufficient for imaging. We usually administer the spasmolytic agent shortly before dynamic imaging and DWI intravenously as balanced steady state free precession images are less prone to motion artefacts and the onset of the effect lies around 1 min. If an intramuscular administration is chosen, patients have to be injected before imaging, as the onset of effect takes about 10 min.

Patients in our department are placed in prone position on the MR table. We use a 50 cm field of view for imaging to cover the whole abdomen from the diaphragm to the symphysis. The prone position reduces the diameter of the abdomen that has to be imaged and thus shortens scan time and also reduces motion artefacts.

3.2.2 Imaging Protocol

Diffusion-weighted imaging can be performed in axial and coronal views. Most authors include one view, but from our own experience, we would always add both orientations, as short pathologies can be missed in one or the other view. Furthermore, coronal views are often more hampered by artefacts [15].

Figure 3.1 shows a typical imaging protocol used in our institution.

The typical slice thickness for DWI lies between 5 and 9 mm [16, 17]. Our protocol includes a coronal DWI sequence with a slice thickness of 7 mm and an axial DWI sequence with a slice thickness of 5 mm.

Diffusion-weighted imaging can be acquired during free breathing, but also breath-hold techniques and respiratory gated sequences are available. We use a free-breathing technique in our study as Kwee et al. showed that DWI data in liver

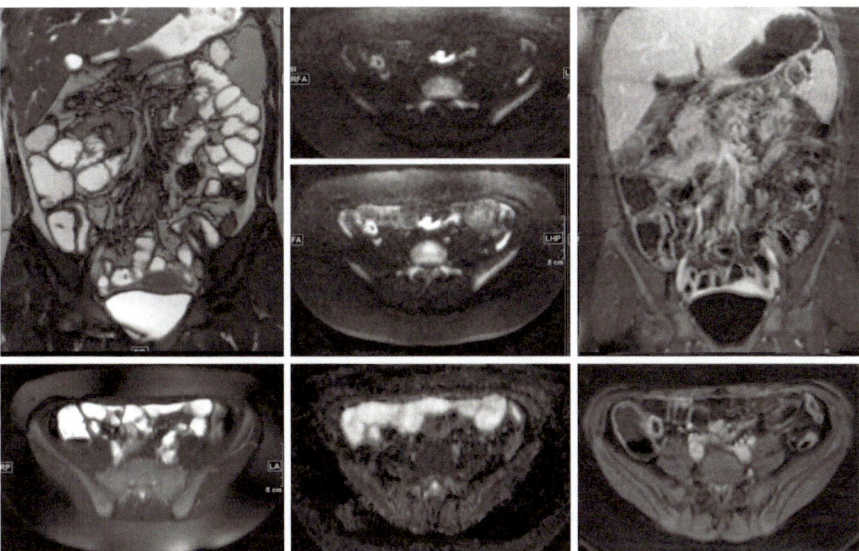

Fig. 3.1 Protocol for small bowel MR imaging used in our clinical routine: left column T2-weighted imaging and TrueFISP; middle column DWI with *b*-values of 50 and 1000 and corresponding ADC map; right column dynamic and post-contrast T1-weighted imaging

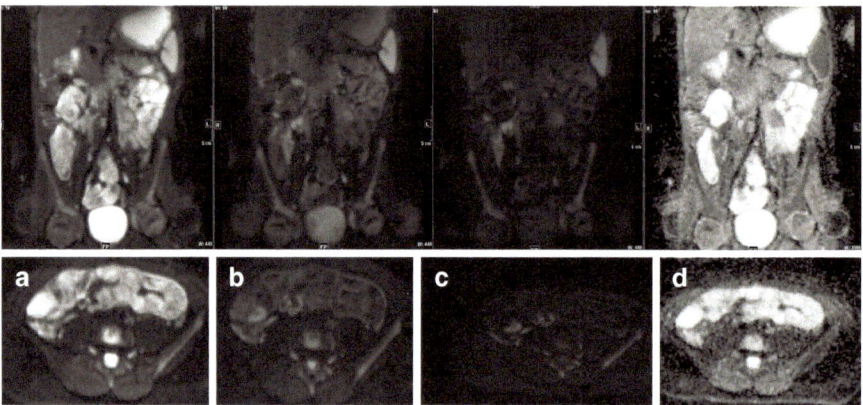

Fig. 3.2 Coronal (upper row) and transverse (lower row) diffusion-weighted imaging (**a–c**) and ADC map (**d**) showing terminal ileitis in a 17-year-old patient. *B*-values of 50 (**a**) show the lesion less good as *b*-values of 500 (**b**) and 1000 (**c**)

imaging were less reproducible when using a breath-hold DWI technique [18]. In a study by Muro et al. in a moving phantom, the authors were able to show that stable motion such as calm respiration does not cause signal loss on DWI [19]. Therefore, DWI using a free-breathing technique is well suited for DWI of the small bowel as patients in prone position mostly do not significantly move on the MR table and bowel movements can be sufficiently suppressed using spasmolytic agents.

Until now, there is no general agreement on which *b*-values should be used for DWI of the small bowel. Most groups perform DWI with two or three *b*-values. The lower *b*-value is chosen as 0 or 50 and reflects a T2-weighted image, where all water-containing structures (especially liquid bowel content) are displayed with high signal intensity. If two *b*-values are chosen, the second one is between 500 and 1000. The higher the *b*-value, the more "normal" fluids are suppressed, and only the tissues with restricted diffusion represent with a higher signal on diffusion-weighted imaging (Fig. 3.2). However, this comes at the cost of lower SNR performance.

We typically use three *b*-values (*b* = 50, 500 and 1000), and from our experience the highest *b*-value shows pathologies in the bowel best most of the time [20].

3.2.3 DWI Analysis

Diffusion-weighted images can be viewed in a qualitative and/or quantitative way. For qualitative viewing images can be analysed as they are or can be colour coded and also superimposed with anatomic T1- or T2-weighted images. The latter resembles hybrid imaging techniques like positron emission tomography (PET)/CT or PET/MR and can help to allocate areas with restricted diffusion to the right anatomic localization more easily than viewing anatomic and DWI images side by side [21]. Furthermore, viewing software provides a wider range of colour contrast compared to greyscale, which can help to differentiate pathologies and allocate them to bowel wall, lumen or extraluminal structures.

Quantitative analysis depends on the acquisition of diffusion-weighted images with different b-values. Apparent diffusion coefficient (ADC) values can be calculated. Regions of interests (ROIs) can be drawn in the bowel wall or other structures to record the specific ADC value of this tissue. Areas of restricted diffusion represent with a low ADC value. On DW images this restricted diffusion represents with a hyperintense signal. ADC value changes under therapy can be used to assess therapy response.

3.3   Inflammatory Bowel Disease

Inflammatory bowel diseases (IBD) represent as recurrent or continuous chronic inflammatory conditions of the bowel due to an autoimmune disease. The two typical and most frequent entities are Crohn's disease (CD) and ulcerative colitis (UC). Other forms of IBD are less frequent. Inflammation in UC is restricted to the mucosa, while CD shows a transmural inflammation. This is why in CD patients extraintestinal abscesses and fistulae can be found. CD can manifest throughout the whole gastrointestinal tract from the mouth to the anus with normal areas in between, whereas UC shows continuous inflammatory lesions in the colon and rectum. The terminal ileum can be involved in both entities but is much more frequently involved in CD: In the majority of CD, the pathology starts in the terminal ileum for which reason the disease is also often called ileitis terminalis. Figures 3.3 and 3.4 show patients with terminal ileitis. In UC the colonic inflammation can sometimes be "washed back" in the upstream terminal ileum and can represent as a so-called backwash ileitis. To focus on small bowel pathologies, the value of diffusion-weighted imaging in CD will be highlighted in the next section.

3.3.1   Crohn's Disease (CD)

Multiple studies have focused on the evaluation of diffusion-weighted imaging in detecting active inflammatory lesions in children and adults. Kiryu et al. and Oto et al. were amongst the first groups reporting on the ability of discriminating inflamed and non-inflamed bowel segments with DWI [17, 22].

Few studies reported higher diagnostic accuracy of DWI compared to gadolinium-enhanced T1-weighted MR imaging [23] and suggested to replace contrast-enhanced MR imaging with DWI. Multiple studies compared standard sequences to protocols with and without DWI: Qi et al. reported that DWI combined with standard sequences has a higher diagnostic accuracy (92%) than standard sequences alone (79%) to detect disease activity [24]. Hahnemann et al. evaluated DWI compared to standard MR imaging in the assessment of inflammatory lesions of the small bowel [25]. They used capsule endoscopy as standard of reference. They found that DWI of the small bowel not only allowed for the detection of inflammatory lesions with

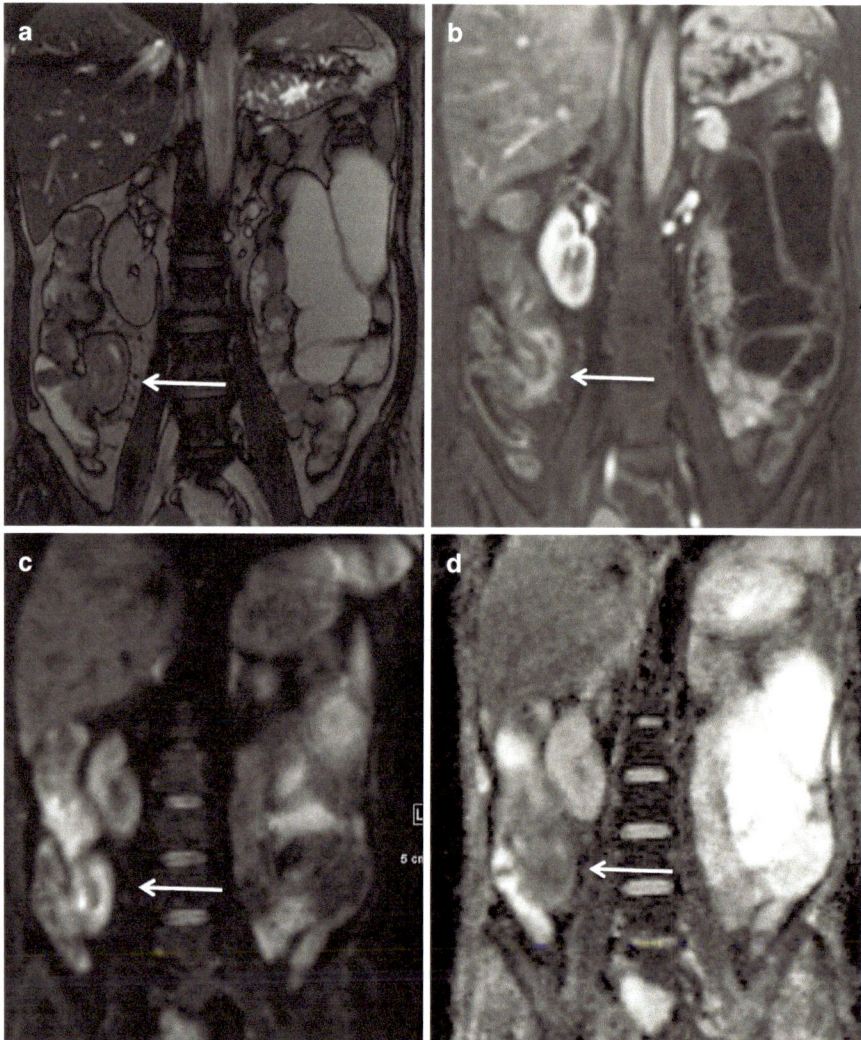

Fig. 3.3 A 15-year-old patient with known Crohn's disease. MR imaging shows terminal ileitis in (**a**) TrueFISP and (**b**) dynamic T1-weighted imaging as well as DWI (**c**, *b*-value 1000; **d**, ADC map)

high accuracy but also enabled the identification of additional lesions that were not found on standard MR imaging alone. Li et al. found similar results when looking at DWI in a paediatric patient population [13]: They found that DWI detected additional inflammatory lesions and reported that in a comprehensive imaging protocol with T2w HASTE and post-contrast scans and DWI, children can be sufficiently examined in <10 min.

Recent studies of Kim et al. and Pendse et al. reported only a small improvement in sensitivity for bowel inflammation by adding DWI to conventional sequences at

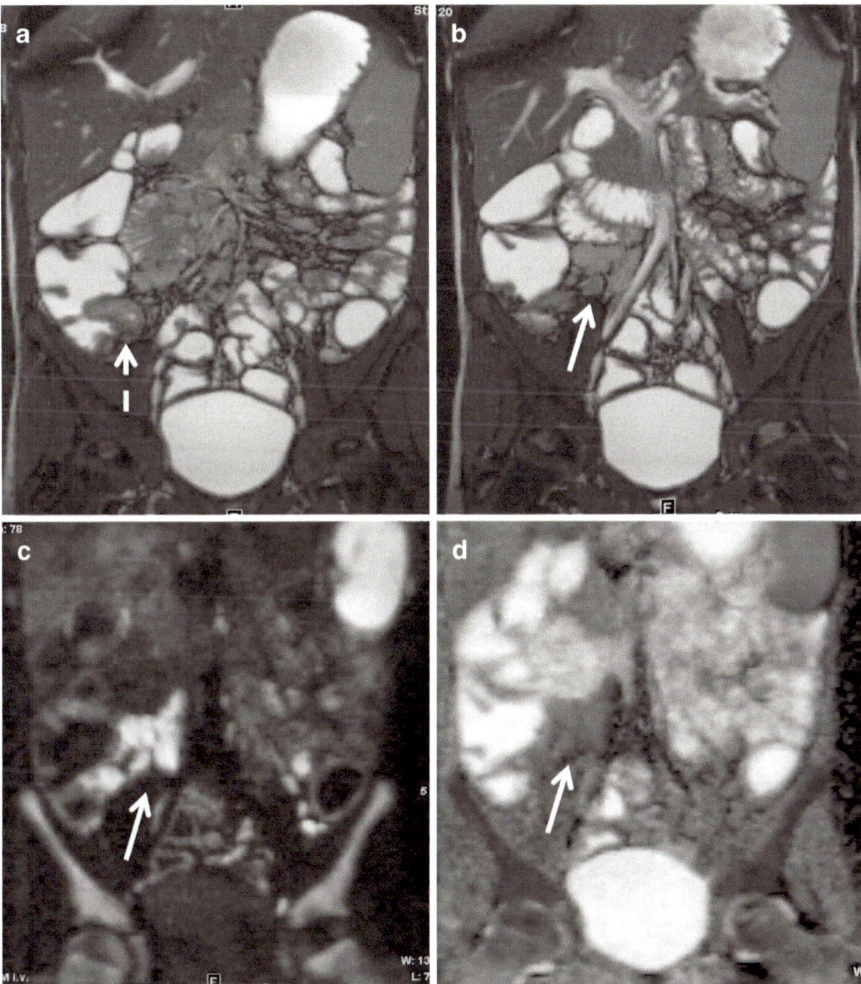

Fig. 3.4 Patient with CD and terminal ileitis (**a**, balanced SSFP with wall thickening of the terminal ileum) showing reactive lymphadenopathy (**b**, TrueFISP; **c**, DWI with *b*-value 500; **d**, ADC map)

the expense of reduced specificity [16, 26]. These study data suggest that DWI cannot replace conventional imaging sequences but can be an adjunct to maintain specificity.

For detection of CD, the initial studies relied on qualitative evaluations of DWI sequences alone. Quantitative evaluations using ADC values have been studied to evaluate response to treatment or to discriminate active disease and chronic fibrosing disease.

Li et al. found a strong correlation between ADC values and a simplified endoscopic score when looking at ileocolonic inflammation [27]. Conventional imaging parameters had a less strong correlation.

Stanescu-Siegmund et al. found an excellent correlation between ADC values and the Harvey Bradshaw Index as clinical parameter [28]. They found an ADC threshold of 1.56×10^{-3} mm^2/s to be able to differentiate between normal and inflamed bowel wall. Similar cut-off values have been reported by Buisson et al. (1.6×10^{-3} mm^2/s) [29], while Zhu et al. reported a cut-off value of 1.11×10^{-3} mm^2/s to have highest sensitivity (100%) and specificity (68.8%).

Kovanlikaya et al. analysed ADC values in paediatric CD patients as a biomarker for fibrosis [30]. They found the ADC values of transmural fibrosis to be lower compared to the reported values of inflammation in Crohn's disease. Tielbeek et al. showed a significant correlation between ADC values and fibrosis in an adult population [31].

Bhatnagar et al. and Buisson et al. examined ADC values (amongst other parameters) and found them useful in detecting and assessing inflammatory activity but also to predict efficacy of anti-TNF induction therapy in CD [32, 33]. Figure 3.5 shows a young female patient with CD before and after anti-inflammatory therapy. Diffusion-weighted imaging depicts the therapy response in a similar way as contrast-enhanced T1-weighted imaging.

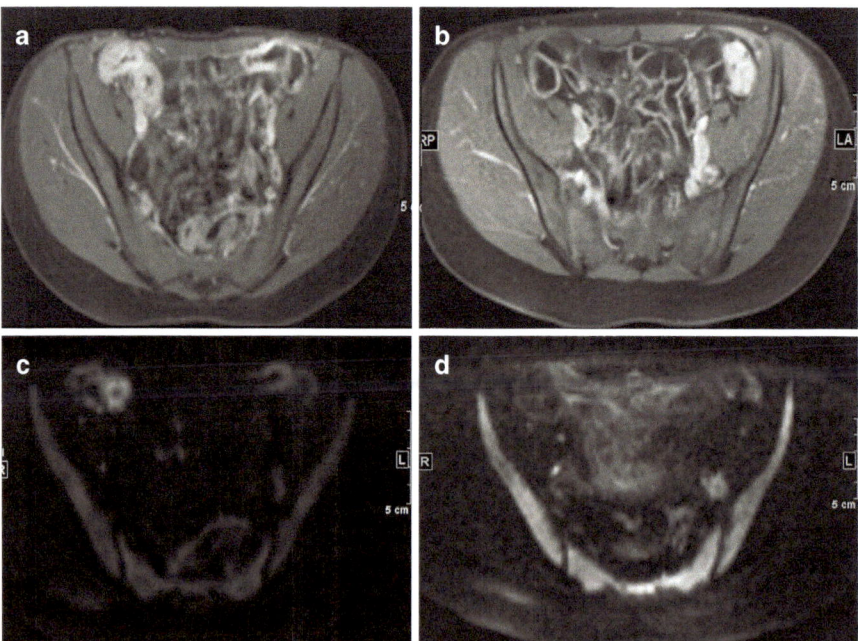

Fig. 3.5 (**a–d**) CD patient before (left) and after (right) therapy: Upper row, contrast-enhanced T1-weighted imaging; lower row, DWI with *b*-value 1000. Before therapy inflammatory lesions in the terminal ileum as well as proximal ileum can be found. These changes are nearly gone in the posttherapeutic control

3.4 Small Bowel Neoplasms

Small bowel neoplasms only account for 1–6% of all gastrointestinal tract tumours [34]. Until now only limited experience on DWI is distributed concerning small bowel tumours [35]. Larger studies have to be performed to establish DWI in small bowel tumour imaging.

3.4.1 Adenocarcinoma

Adenocarcinomas are the most frequent malignancies of the small bowel, accounting for about 40%. They usually appear in the proximal small bowel and present as an annular lesion showing T2 hypointensity frequently and a heterogenous moderate enhancement on gadolinium-enhanced T1-weighted images [36]. On diffusion-weighted imaging, tumours show a diffusion restriction. By now, there is no study reporting explicitly on ADC values for adenocarcinoma. Amzallag-Bellenger et al. have reported on three patients with adenocarcinoma of the jejunum or ileum, who presented with miscellaneous ADC values measuring between 0.76 and 1.28×10^{-3} mm^2/s.

3.4.2 Lymphoma

Lymphomas account for about 20% of all small bowel tumours. They can be found mostly in the distal ileum due to the large amount of lymphoid tissue at this site. Patients suffering from Crohn's disease, celiac disease and extraintestinal lymphoma and those who are immunocompromised or have had chemotherapy have a higher risk to develop small bowel lymphoma. Typically, lymphoma presents as a thickened mass of the wall without complete obstruction showing considerable dilatation. Polypoid lesions with protrusion into the bowel lumen or eccentric masses, which can also show mural ulcerations and fistulation, are much less common. Infiltration of the mesenteric fat in the absence of discrete lymphadenopathy seems to be associated with high-grade non-Hodgkin lymphoma [37].

On diffusion-weighted imaging, lymphomas have been reported to show a mean ADC value of $0.66 \pm 0.19 \times 10^{-3}$ mm^2/s [35].

Figure 3.6 shows a 13-year-old boy diagnosed with high-grade lymphoma of the duodenum. Diffusion restriction is obvious in the lesion and ADC value clearly low.

3.4.3 Carcinoids

Carcinoids account for about 2% of GI tumours. In the small bowel, carcinoids are the second most common tumours (31%). The most common localization for carcinoids is the appendix; second most commonly they can be found in the distal ileum.

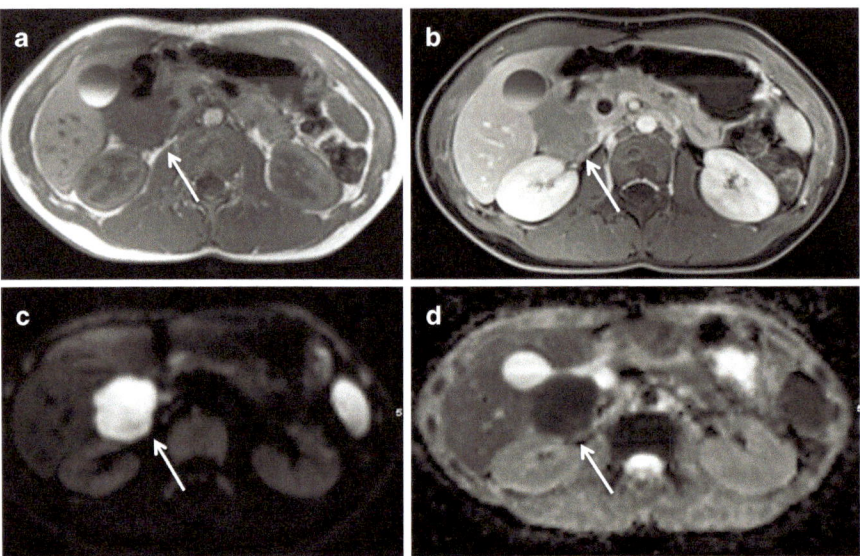

Fig. 3.6 (**a–d**) A 14-year-old boy with diagnosis of a high-grade lymphoma in the duodenum. Diffusion-weighted imaging (lower row) shows the diffusion restriction, resulting in a low signal on ADC map

In about 30% of cases, carcinoids present in multiple locations. Metastases to the lymph nodes and liver occur depending on the size of the bowel lesion. On MR imaging carcinoids present as mostly isointense lesions on unenhanced T1- and T2-weighted images. Especially the tumours that occur in the distal small bowel often show mesenteric masses with similar imaging features, which are 2–4 cm in size. Ileal carcinoids often present as a mesenteric mass with radiating strands of tissue [36, 38]. Annular narrowing is rarely seen with carcinoids, but the involvement of the adjacent mesentery stimulates desmoplastic reactions and fibrosis and can result in angulation and kinking of the bowel leading to obstruction and ischemia.

On diffusion-weighted imaging, carcinoids have been reported to show a mean ADC value of $0.83 \pm 0.29 \times 10^{-3}$ mm^2/s [35].

3.4.4 Gastrointestinal Stromal Tumours (GISTs)

GISTs are the most common mesenchymal tumours of the GI tract and can be accounted for 9% of all tumours of the GI tract. Most frequently they can be found in the stomach (up to 60%), followed by the jejunum and ileum (30%) and less so in the duodenum (5%). They can be small or large lesions. In the small bowel, small GISTs usually are round and present with a strong homogeneous arterial enhancement. Larger GISTs tend to present as lobulated lesions with less enhancement and cystic changes [39]. Yu et al. were able to show that the mean ADC value measured in the GISTs and the malignancy risk of the tumour correlated negatively. Diffusion-weighted imaging in GISTs has been reported to show a mean ADC value of

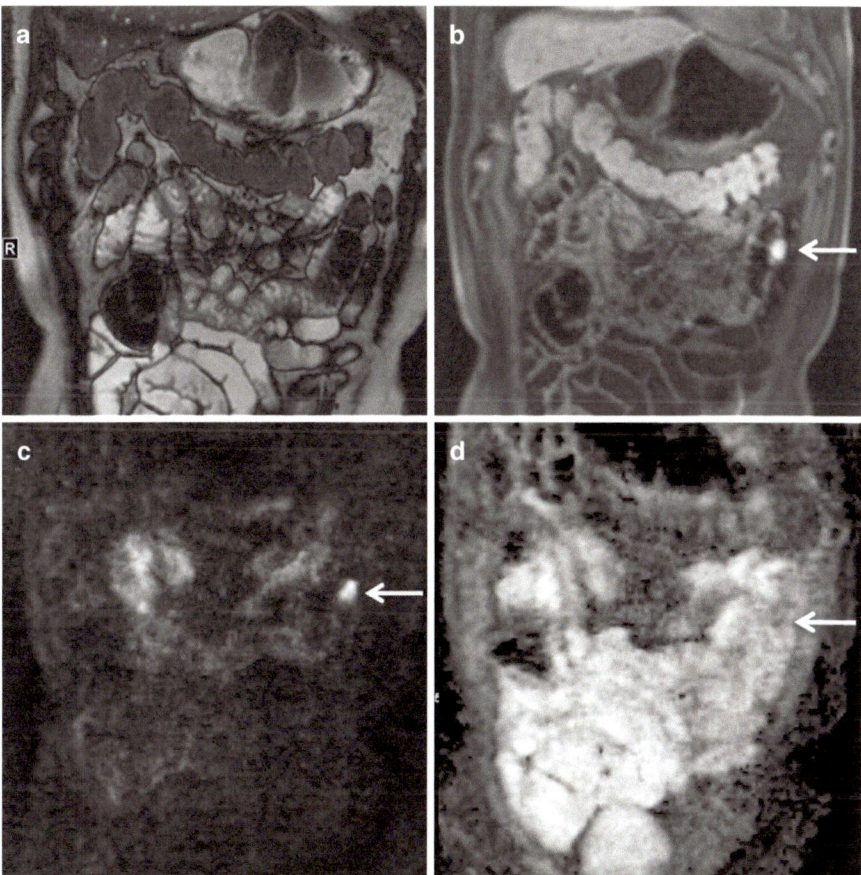

Fig. 3.7 A 72-year-old male patient with a small gastrointestinal stromal tumour (GIST) in the proximal ileum. High signal on TrueFISP imaging (**a**) and typical homogenous enhancement on arterial dynamic T1-weighted imaging after contrast injection (**b**). DWI (**c**) and ADC map (**d**) show the restricted diffusion, but less low ADC values

$1.24 \pm 0.34 \times 10^{-3}$ mm²/s [35]. Figure 3.7 shows a 72-year-old patient with a small GIST of the proximal ileum with classical homogenous arterial enhancement. Compared to the lymphoma shown in Fig. 3.6, ADC value is not as low as in the lymphoma, but diffusion restriction is also clearly shown.

3.5   Other Small Bowel Pathologies

Amongst the wide spectrum of small bowel pathologies, three very important and barely rare entities are presented in this chapter part. For all mentioned small bowel pathologies, patient's clinical presentation and laboratory markers are essential to help differentiate based on their MR imaging changes including diffusion characteristics.

3.5.1 Gluten-Sensitive Enteropathy

Gluten-sensitive enteropathy also known as celiac disease or celiac sprue is a T-cell-mediated disease occurring in genetically susceptible individuals induced by the ingestion of one of several proteins found in wheat (gliadins), barley (hordeins) and rye (secalinin) [40]. With a prevalence of up to 2%, it is the most frequent enteropathy in Western countries. This food can induce inflammatory processes, which can then destroy the bowel epithelial lining. Symptoms and severity of the disease can be different, and therefore MR imaging not always shows the pathology. If an inflammatory process is present, DWI can help to distinguish non-diseased from diseased bowel parts. Figure 3.8 shows a patient with known gluten-sensitive enteropathy and an inflammatory process in the small bowel.

3.5.2 Vasculitis

Vasculitis occurs as a systemic disease, e.g. in lupus erythematosus, polyarteritis nodosa and Henoch-Schönlein purpura. Vasculitis affects the distal small vessels, which are usually occluded or narrowed resulting in local ischemia. Usually, a relatively long bowel part is involved and the distribution is non-segmental. Involvement of the duodenum by ischemic changes is nearly always indicative of vasculitis [41]. Figure 3.9 shows a patient with ANCA (anti-neutrophil cytoplasmic autoantibody) vasculitis who was initially diagnosed on computed tomography. The follow-up examination under therapy was performed with MR imaging where changes of the small bowel caused by vasculitis can still be seen. Diffusion-weighted imaging shows the restricted diffusion due to inflammatory changes in the bowel wall.

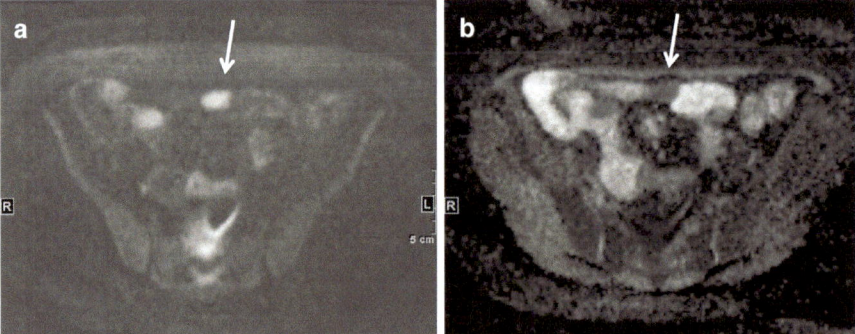

Fig. 3.8 A 43-year-old female patient with known gluten-sensitive enteropathy and an inflammatory process in the small bowel presenting as restricted diffusion on DWI (**a**) and low signal on ADC map (**b**)

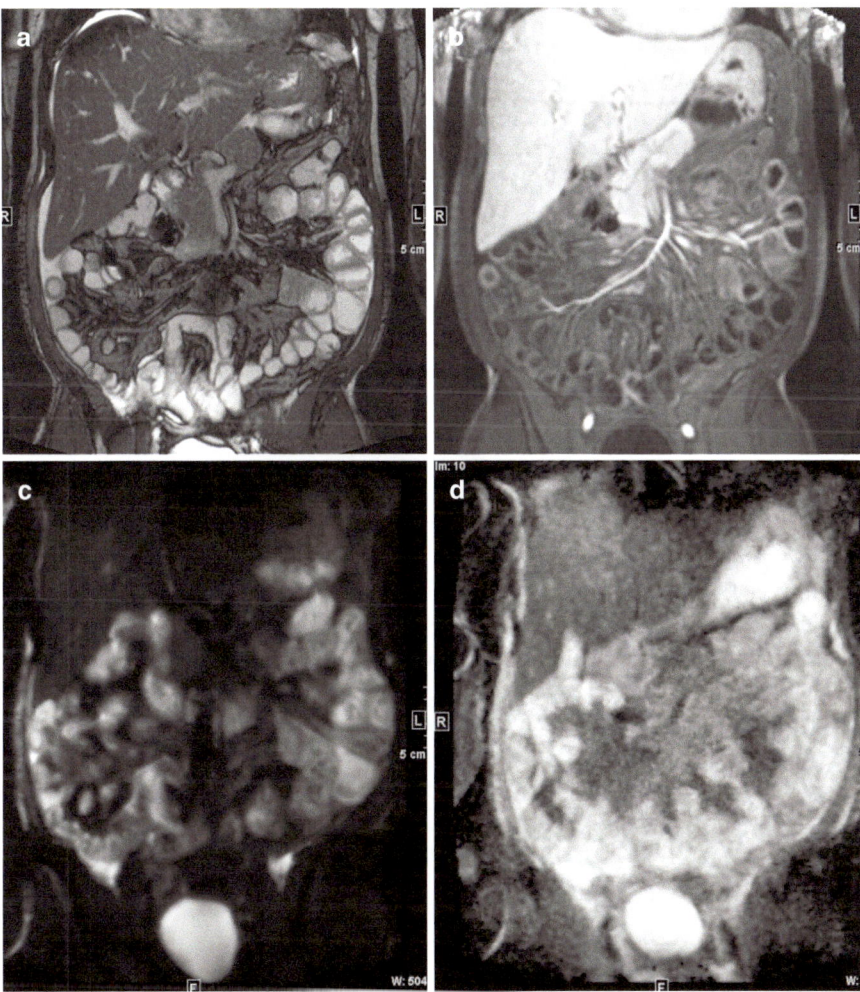

Fig. 3.9 Patient with ANCA vasculitis, already under therapy. (**a**) balanced SSFP; (**b**) arterial dynamic phase; (**c**) DWI with a *b*-value of 1000; (**d**) ADC map still show multiple affected small bowel loops in the left upper and right lower abdomen

3.5.3   Therapy-Induced Changes of the Small Bowel

Chemotherapy and radiation can alter the small bowel, and these changes can be determined on MR enterography. Subacute changes after radiation therapy (5–12 months after completion) can be ascribed to severe endarteritis obliterans with resultant inflammatory bowel changes which can be seen on DWI as well. Chemotherapy can lead to focal or diffuse small bowel changes.

Another entity that occurs frequently after therapy is graft-versus-host disease (GvHD) of the bowel in patients after stem cell transplantation (SCT). GvHD is one of the major causes of mortality and morbidity after allogenic SCT. It develops frequently after SCT (typically 3–11 weeks after transplantation), and 30–50% of patients are affected. Intestinal manifestation is one of the most frequent. The findings are similar to those in CD, but the extent is usually greater [42]. Diffusion-weighted imaging can help to show diseased bowel parts. Figure 3.10 shows a patient with graft-versus-host disease after stem cell transplantation for acute myeloic leucemia.

With the knowledge of previous therapy in the future, it might be possible to omit contrast injection in this patient as DWI is able to show the changes just quite as well as contrast-enhanced T1-weighted sequences.

3.6   Appendicitis

Acute appendicitis is a common medical emergency condition that can affect children and adults. The incidence of appendicitis has been shown to increase, and the lifetime risk lies around 9%. Clinical symptoms and laboratory parameters alone often render the diagnosis of appendicitis. Ultrasound has been the diagnostic method of choice to evaluate for appendicitis but is dependent on the system used, the operator as well as the patient, especially if patients are larger [43] or if the appendix lies in a not well-visible area, e.g. retrocaecal. CT can visualize the location and condition of the appendix well but is burdened by use of ionizing radiation, which is especially a problem in paediatric and pregnant patients. The increasing prevalence and accessibility of MR have led many institutions to choose MR as the primary cross-sectional imaging tool for appendicitis. MR imaging combines the advantages of ultrasound (noninvasive, lack of ionizing radiation) with the high-resolution 3D cross-sectional information of CT [44].

Standard imaging sequences alone have shown high specificity and sensitivity in diagnosing acute appendicitis [45, 46]. Diffusion-weighted imaging has been proven to add important information [47]. Acute appendicitis appears bright on fat-suppressed DWI due to the combination of restricted diffusion and oedema that appears bright with T2 weighting. As with small bowel imaging, we typically use two to three b-values for appendicitis visualization, most often $b = 0$ and $b = 500$. The $b = 0$ image is effectively a fat-suppressed T2-weighted image and is ideal for visualizing periappendiceal fluids. We have found that $b = 500$ is a good compromise for imaging of acute appendicitis, balancing adequate diffusion weighting and good SNR performance [47]. In general, we have not found the quantitative ADC maps particularly useful for the diagnosis of acute appendicitis, although we have found the high b-value images to be very helpful for qualitative detection of oedema and inflammatory changes. Finally, we routinely use externally calibrated parallel imaging with all DWI acquisitions to reduce distortion in the phase encoding direction due to the sensitivity of echo-planar methods to magnetic field inhomogeneities.

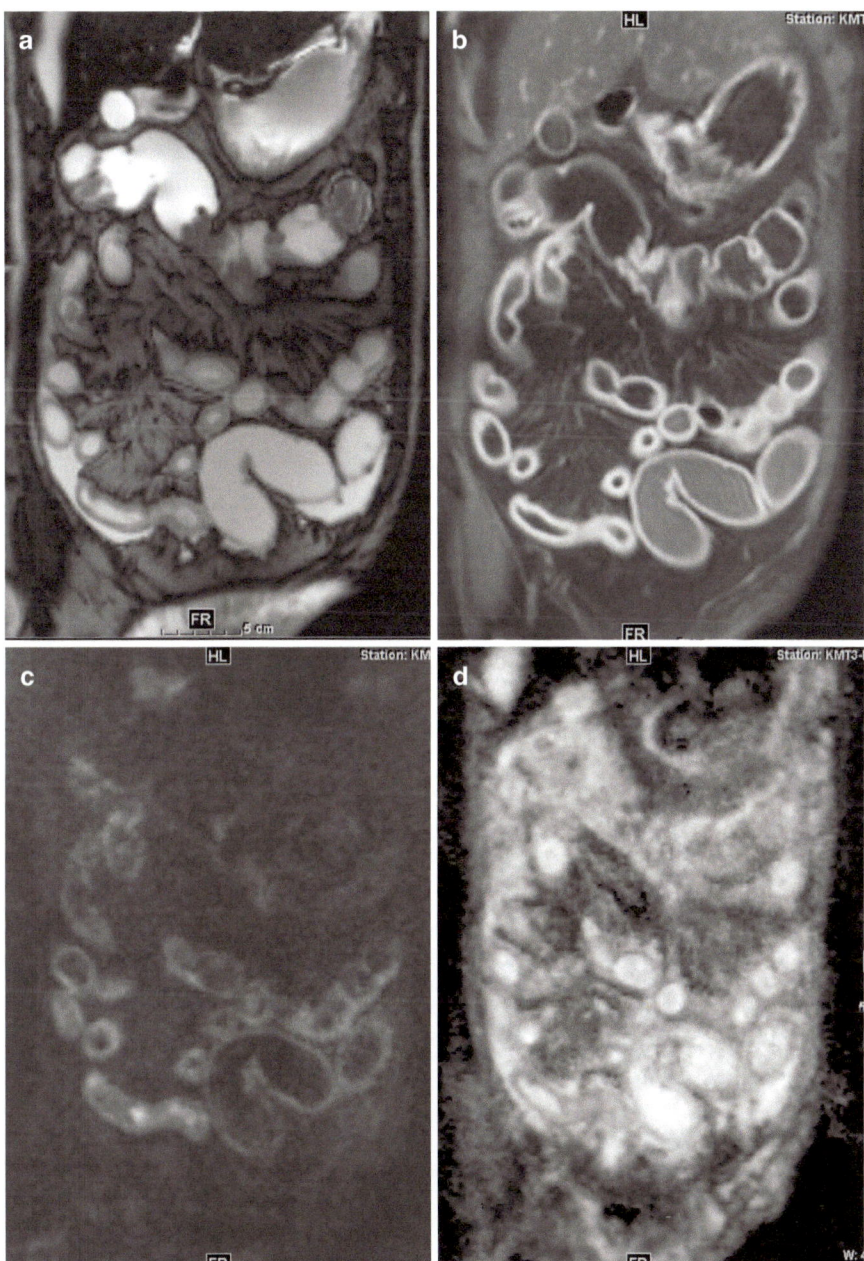

Fig. 3.10 Patient with graft-versus-host disease after stem cell transplantation for acute myeloic leucemia. Small bowel and stomach are affected as shown by bowel wall thickening in TrueFISP (**a**), contrast enhancement in T1-weighted imaging after contrast injection (**b**) and DWI (**c**, *b*-value 500; **d**, ADC map)

Studies have shown the value of adding DWI to a routine imaging protocol to diagnose acute appendicitis: Bayraktutan et al. studied 45 consecutive children suspected of having appendicitis and compared the diagnostic performance of DWI with standard sequences and surgical findings [48]: A combination of DWI and conventional MR imaging showed highest sensitivity and specificity compared to standard sequences and DWI alone. They found mean ADC values for inflamed appendices to be $1.12 \pm 0.17 \times 10^{-3}$ mm^2/s, while normal appendices showed a mean ADC value of $2.17 \times 0.11 \times 10^{-3}$ mm^2/s. This is in accordance with results found in adults: Inci et al. examined 119 patients with a suspicion of acute appendicitis and 50 control patients [49]. They found mean ADC values in healthy appendices to be $2.02 \pm 0.19 \times 10^{-3}$ mm^2/s, and in inflamed appendices the mean ADC value was $1.22 \pm 0.18 \times 10^{-3}$ mm^2/s. Like us, they found a b-value of 500 to be of highest value for visualizing appendiceal inflammation, even if a higher b-value of 1000 was also used.

Figure 3.11 shows an image example of a patient diagnosed with appendicitis who went straight to appendectomy after the examination: diffusion-weighted imaging is able to show the inflammation as well as contrast-enhanced MR imaging. As not all patients undergo surgery for inflammatory appendiceal masses (IAM), DWI was studied as a follow-up tool in appendicitis by Özdemir et al. [50]: they concluded that DWI may be used with a significant success for follow-up of patients with IAM. As a monitoring imaging method, DWI may also aid in determining the most appropriate timing for interval appendectomy as well as help in diagnosing alternative diagnoses (e.g. malignancy and inflammatory bowel disease) that can mimic IAM.

As DWI is a quite new tool to diagnose appendicitis, most radiologists have limited or no experience in the evaluation and most likely will need training to achieve the diagnostic accuracy that has been reported in the literature [51].

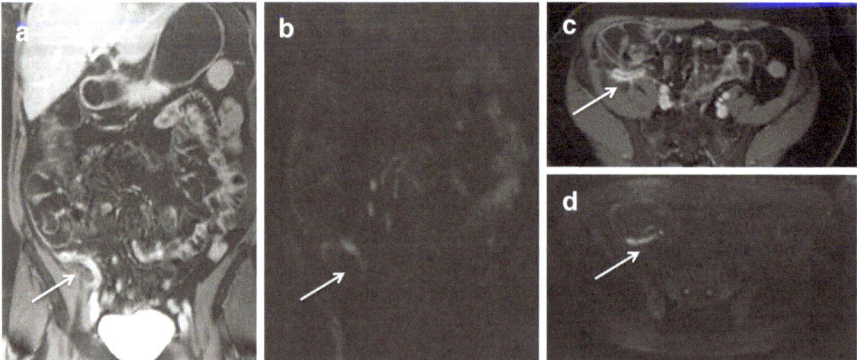

Fig. 3.11 A patient presenting to the emergency room with a high suspicion of appendicitis and an unequivocal ultrasound. Contrast-enhanced T1-weighted images (**a**, coronal; **c**, axial) show contrast enhancement as correlate for acute inflammation. Diffusion-weighted imaging (**b**, coronal; **d**, axial) shows high signal as result of restricted diffusion due to inflammation

3.7 Summary

Diffusion-weighted imaging has proven to be of help in the imaging of the small bowel and appendix. While conventional imaging sequences still are the "backbone" of small bowel imaging, DWI can add quantitative and qualitative information. DWI therefore should be part of a routine small bowel protocol to diagnose appendicitis in MR imaging. Standard MR imaging sequences including T2-weighted images with and without fat suppression as well as contrast-enhanced imaging will continue to be the reference standards until larger studies have been performed, especially in patients with small bowel tumours.

References

1. Rondonotti E, Herrerias JM, Pennazio M, Caunedo A, Mascarenhas-Saraiva M, de Franchis R. Complications, limitations, and failures of capsule endoscopy: a review of 733 cases. Gastrointest Endosc. 2005;62(5):712–6; . quiz 52, 54. https://doi.org/10.1016/j.gie.2005.05.002.
2. Sinha R, Rajiah P, Ramachandran I, Sanders S, Murphy PD. Diffusion-weighted MR imaging of the gastrointestinal tract: technique, indications, and imaging findings. Radiographics. 2013;33(3):655–76; . discussion 76–80. https://doi.org/10.1148/rg.333125042.
3. Dillman JR, Smith EA, Sanchez R, Adler J, Fazeli S, Zhang B, et al. DWI in pediatric small-bowel Crohn disease: are apparent diffusion coefficients surrogates for disease activity in patients receiving infliximab therapy? AJR Am J Roentgenol. 2016;207(5):1002–8. https://doi.org/10.2214/AJR.16.16477.
4. Kanda T, Ishii K, Kawaguchi H, Kitajima K, Takenaka D. High signal intensity in the dentate nucleus and globus pallidus on unenhanced T1-weighted MR images: relationship with increasing cumulative dose of a gadolinium-based contrast material. Radiology. 2014;270(3):834–41. https://doi.org/10.1148/radiol.13131669.
5. Umschaden HW, Gasser J. MR enteroclysis. Radiol Clin North Am. 2003;41(2):231–48.
6. Masselli G, Casciani E, Polettini E, Gualdi G. Comparison of MR enteroclysis with MR enterography and conventional enteroclysis in patients with Crohn's disease. Eur Radiol. 2008;18(3):438–47. https://doi.org/10.1007/s00330-007-0763-2.
7. Arrivé L, El Mouhadi S. MR enterography versus MR enteroclysis. Radiology. 2013;266(2):688. https://doi.org/10.1148/radiol.12121840.
8. Negaard A, Paulsen V, Sandvik L, Berstad A, Borthne A, Try K, et al. A prospective randomized comparison between two MRI studies of the small bowel in Crohn's disease, the oral contrast method and MR enteroclysis. Eur Radiol. 2007;17(9):2294–301. https://doi.org/10.1007/s00330-007-0648-4.
9. Schreyer AG, Geissler A, Albrich H, Schölmerich J, Feuerbach S, Rogler G, et al. Abdominal MRI after enteroclysis or with oral contrast in patients with suspected or proven Crohn's disease. Clin Gastroenterol Hepatol. 2(6):491–7. https://doi.org/10.1016/S1542-3565(04)00168-5.
10. Lauenstein TC, Umutlu L, Kloeters C, Aschoff AJ, Ladd ME, Kinner S. Small bowel imaging with MRI. Acad Radiol. 2012;19(11):1424–33. https://doi.org/10.1016/j.acra.2012.05.019.
11. Young BM, Fletcher JG, Booya F, Paulsen S, Fidler J, Johnson CD, et al. Head-to-head comparison of oral contrast agents for cross-sectional enterography: small bowel distention, timing, and side effects. J Comput Assist Tomogr. 2008;32(1):32–8. https://doi.org/10.1097/RCT.0b013e318061961d.
12. Oussalah A, Laurent V, Bruot O, Bressenot A, Bigard M-A, Régent D, et al. Diffusion-weighted magnetic resonance without bowel preparation for detecting colonic inflammation in inflammatory bowel disease. Gut. 2010;59(8):1056–65. https://doi.org/10.1136/gut.2009.197665.

13. Li M, Dick A, Hassold N, Pabst T, Bley T, Kostler H, et al. CAIPIRINHA-accelerated T1w 3D-FLASH for small-bowel MR imaging in pediatric patients with Crohn's disease: assessment of image quality and diagnostic performance. World J Pediatr. 2016;12(4):455–62. https://doi.org/10.1007/s12519-016-0047-5.

14. Froehlich JM, Daenzer M, von Weymarn C, Erturk SM, Zollikofer CL, Patak MA. Aperistaltic effect of hyoscine N-butylbromide versus glucagon on the small bowel assessed by magnetic resonance imaging. Eur Radiol. 2009;19(6):1387–93. https://doi.org/10.1007/s00330-008-1293-2.

15. Sirin S, Kathemann S, Schweiger B, Hahnemann ML, Forsting M, Lauenstein TC, et al. Magnetic resonance colonography including diffusion-weighted imaging in children and adolescents with inflammatory bowel disease: do we really need intravenous contrast? Invest Radiol. 2015;50(1):32–9. https://doi.org/10.1097/RLI.0000000000000092.

16. Pendse DA, Makanyanga JC, Plumb AA, Bhatnagar G, Atkinson D, Rodriguez-Justo M, et al. Diffusion-weighted imaging for evaluating inflammatory activity in Crohn's disease: comparison with histopathology, conventional MRI activity scores, and faecal calprotectin. Abdom Radiol. 2017;42(1):115–23. https://doi.org/10.1007/s00261-016-0863-z.

17. Oto A, Kayhan A, Williams JTB, Fan X, Yun L, Arkani S, et al. Active Crohn's Disease in the small bowel: evaluation by diffusion weighted imaging and quantitative dynamic contrast enhanced MR imaging. J Magn Reson Imaging. 2011;33(3):615–24. https://doi.org/10.1002/jmri.22435.

18. Kwee TC, Takahara T, Koh D-M, Nievelstein RAJ, Luijten PR. Comparison and reproducibility of ADC measurements in breathhold, respiratory triggered, and free-breathing diffusion-weighted MR imaging of the liver. J Magn Reson Imaging. 2008;28(5):1141–8. https://doi.org/10.1002/jmri.21569.

19. Muro I, Takahara T, Horie T, Honda M, Kamiya A, Okumura Y et al. [Influence of respiratory motion in body diffusion weighted imaging under free breathing (examination of a moving phantom)]. Nihon Hoshasen Gijutsu Gakkai Zasshi 2005;61(11):1551–1558.

20. van Rijswijck C, Lauenstein T, Kinner S. Detectability of inflammatory bowel disease in Diffusion-weighted MR imaging (DWI): which imaging plane and b-values should be preferred? Insights Imaging. 2017:262.

21. Ehman EC, Phelps AS, Ohliger MA, Rhee SJ, MacKenzie JD, Courtier JL. Detection of bowel inflammation with fused DWI/T2 images versus contrast-enhanced images in pediatric MR enterography with histopathologic correlation. Clin Imaging. 2016;40(6):1135–9. https://doi.org/10.1016/j.clinimag.2016.07.006.

22. Kiryu S, Dodanuki K, Takao H, Watanabe M, Inoue Y, Takazoe M, et al. Free-breathing diffusion-weighted imaging for the assessment of inflammatory activity in Crohn's disease. J Magn Reson Imaging. 2009;29(4):880–6. https://doi.org/10.1002/jmri.21725.

23. Dubron C, Avni F, Boutry N, Turck D, Duhamel A, Amzallag-Bellenger E. Prospective evaluation of free-breathing diffusion-weighted imaging for the detection of inflammatory bowel disease with MR enterography in childhood population. Br J Radiol. 2016;89(1060):20150840. https://doi.org/10.1259/bjr.20150840.

24. Qi F, Jun S, Qi QY, Chen PJ, Chuan GX, Jiong Z, et al. Utility of the diffusion-weighted imaging for activity evaluation in Crohn's disease patients underwent magnetic resonance enterography. BMC Gastroenterol. 2015;15:12. https://doi.org/10.1186/s12876-015-0235-0.

25. Hahnemann ML, Dechene A, Kathemann S, Sirin S, Gerken G, Lauenstein TC, et al. Diagnostic value of diffusion-weighted imaging (DWI) for the assessment of the small bowel in patients with inflammatory bowel disease. Clin Radiol. 2017;72(1):95.e1–8. https://doi.org/10.1016/j.crad.2016.08.007.

26. Kim KJ, Lee Y, Park SH, Kang BK, Seo N, Yang SK, et al. Diffusion-weighted MR enterography for evaluating Crohn's disease: how does it add diagnostically to conventional MR enterography? Inflamm Bowel Dis. 2015;21(1):101–9. https://doi.org/10.1097/MIB.0000000000000222.

27. Li XH, Sun CH, Mao R, Huang SY, Zhang ZW, Yang XF, et al. Diffusion-weighted MRI enables to accurately grade inflammatory activity in patients of ileocolonic Crohn's disease:

results from an observational study. Inflamm Bowel Dis. 2017;23(2):244–53. https://doi.org/10.1097/MIB.0000000000001001.

28. Stanescu-Siegmund N, Nimsch Y, Wunderlich AP, Wagner M, Meier R, Juchems MS, et al. Quantification of inflammatory activity in patients with Crohn's disease using diffusion weighted imaging (DWI) in MR enteroclysis and MR enterography. Acta Radiol. 2017;58(3):264–71. https://doi.org/10.1177/0284185116648503.

29. Buisson A, Joubert A, Montoriol PF, Ines DD, Hordonneau C, Pereira B, et al. Diffusion-weighted magnetic resonance imaging for detecting and assessing ileal inflammation in Crohn's disease. Aliment Pharmacol Ther. 2013;37(5):537–45. https://doi.org/10.1111/apt.12201.

30. Kovanlikaya A, Dencck D, Rose M, Renjen P, Dunning A, Solomon A, et al. Quantitative apparent diffusion coefficient (ADC) values as an imaging biomarker for fibrosis in pediatric Crohn's disease: preliminary experience. Abdom Imaging. 2015;40(5):1068–74. https://doi.org/10.1007/s00261-014-0247-1.

31. Tielbeek JA, Ziech ML, Li Z, Lavini C, Bipat S, Bemelman WA, et al. Evaluation of conventional, dynamic contrast enhanced and diffusion weighted MRI for quantitative Crohn's disease assessment with histopathology of surgical specimens. Eur Radiol. 2014;24(3):619–29. https://doi.org/10.1007/s00330-013-3015-7.

32. Bhatnagar G, Dikaios N, Prezzi D, Vega R, Halligan S, Taylor SA. Changes in dynamic contrast-enhanced pharmacokinetic and diffusion-weighted imaging parameters reflect response to anti-TNF therapy in Crohn's disease. Br J Radiol. 2015;88(1055):20150547. https://doi.org/10.1259/bjr.20150547.

33. Buisson A, Hordonneau C, Goutte M, Scanzi J, Goutorbe F, Klotz T, et al. Diffusion-weighted magnetic resonance enterocolonography in predicting remission after anti-TNF induction therapy in Crohn's disease. Dig Liver Dis. 2016;48(3):260–6. https://doi.org/10.1016/j.dld.2015.10.019.

34. Gourtsoyiannis N, Papanikolaou N. MR enteroclysis. In: Gore R, Levine M, editors. Textbook of gastrointestinal radiology. Philadelphia, PA: Saunders Elsevier; 2008. p. 765–74.

35. Amzallag-Bellenger E, Soyer P, Barbe C, Nguyen TL, Amara N, Hoeffel C. Diffusion-weighted imaging for the detection of mesenteric small bowel tumours with Magnetic Resonance-enterography. Eur Radiol. 2014;24(11):2916–26. https://doi.org/10.1007/s00330-014-3303-x.

36. Semelka RC, John G, Kelekis NL, Burdeny DA, Ascher SM. Small bowel neoplastic disease: demonstration by MRI. J Magn Reson Imaging. 1996;6(6):855–60.

37. Lohan DG, Alhajeri AN, Cronin CG, Roche CJ, Murphy JM. MR enterography of small-bowel lymphoma: potential for suggestion of histologic subtype and the presence of underlying celiac disease. AJR Am J Roentgenol. 2008;190(2):287–93. https://doi.org/10.2214/AJR.07.2721.

38. Masselli G, Gualdi G. Evaluation of small bowel tumors: MR enteroclysis. Abdom Imaging. 2010;35(1):23–30. https://doi.org/10.1007/s00261-008-9490-7.

39. Yu MH, Lee JM, Baek JH, Han JK, Choi BI. MRI features of gastrointestinal stromal tumors. AJR Am J Roentgenol. 2014;203(5):980–91. https://doi.org/10.2214/AJR.13.11667.

40. Gheller-Rigoni AI, Yale SH, Abdulkarim AS. Celiac Disease: celiac sprue, gluten-sensitive enteropathy. Clin Med Res. 2004;2(1):71–2.

41. Rha SE, Ha HK, Lee SH, Kim JH, Kim JK, Kim JH, et al. CT and MR imaging findings of bowel ischemia from various primary causes. Radiographics. 2000;20(1):29–42. https://doi.org/10.1148/radiographics.20.1.g00ja0629.

42. Derlin T, Laqmani A, Veldhoen S, Apostolova I, Ayuk F, Adam G, et al. Magnetic resonance enterography for assessment of intestinal graft-versus-host disease after allogeneic stem cell transplantation. Eur Radiol. 2015;25(5):1229–37. https://doi.org/10.1007/s00330-014-3503-4.

43. Pickuth D, Heywang-Kobrunner SH, Spielmann RP. Suspected acute appendicitis: is ultrasonography or computed tomography the preferred imaging technique? Eur J Surg. 2000;166(4):315–9. https://doi.org/10.1080/110241500750009177.

44. Liu B, Ramalho M, AlObaidy M, Busireddy KK, Altun E, Kalubowila J, et al. Gastrointestinal imaging-practical magnetic resonance imaging approach. World J Radiol. 2014;6(8):544–66. https://doi.org/10.4329/wjr.v6.i8.544.

45. Repplinger MD, Levy JF, Peethumnongsin E, Gussick ME, Svenson JE, Golden SK, et al. Systematic review and meta-analysis of the accuracy of MRI to diagnose appendicitis in the general population. J Magn Reson Imaging. 2016;43(6):1346–54. https://doi.org/10.1002/jmri.25115.
46. Heverhagen JT, Pfestroff K, Heverhagen AE, Klose KJ, Kessler K, Sitter H. Diagnostic accuracy of magnetic resonance imaging: a prospective evaluation of patients with suspected appendicitis (diamond). J Magn Reson Imaging. 2012;35(3):617–23. https://doi.org/10.1002/jmri.22854.
47. Kinner S, Repplinger MD, Pickhardt PJ, Reeder SB. Contrast-enhanced abdominal MRI for suspected appendicitis: how we do it. AJR Am J Roentgenol. 2016;207(1):49–57. https://doi.org/10.2214/AJR.15.15948.
48. Bayraktutan U, Oral A, Kantarci M, Demir M, Ogul H, Yalcin A, et al. Diagnostic performance of diffusion-weighted MR imaging in detecting acute appendicitis in children: comparison with conventional MRI and surgical findings. J Magn Reson Imaging. 2014;39(6):1518–24. https://doi.org/10.1002/jmri.24316.
49. Inci E, Kilickesmez O, Hocaoglu E, Aydin S, Bayramoglu S, Cimilli T. Utility of diffusion-weighted imaging in the diagnosis of acute appendicitis. Eur Radiol. 2011;21(4):768–75. https://doi.org/10.1007/s00330-010-1981-6.
50. Özdemir O, Metin Y, Metin NO, Küpeli A, Kalcan S, Taşçı F. Contribution of diffusion-weighted MR imaging in follow-up of inflammatory appendiceal mass: preliminary results and review of the literature. Eur J Radiol Open. 2016;3:207–15. https://doi.org/10.1016/j.ejro.2016.08.005.
51. Leeuwenburgh MM, Wiarda BM, Bipat S, Nio CY, Bollen TL, Kardux JJ, et al. Acute appendicitis on abdominal MR images: training readers to improve diagnostic accuracy. Radiology. 2012;264(2):455–63. https://doi.org/10.1148/radiol.12111896.

Large Bowel

4

Luís Curvo Semedo

4.1   Introduction

Magnetic resonance imaging (MRI) is being increasingly performed for the assessment of lesions of the large bowel, whereby mainly morphometric macroscopic tissue information is usually obtained. Yielding insights at a cellular level, diffusion-weighted imaging (DWI) provides images whose signal intensity is sensitized to the random motion of free water molecules. The mobility of water molecules within a given voxel is determined by the microscopic cellular structure, i.e., the presence of barriers, such as cell membranes and macromolecules. Thus, DWI offers a theoretical possibility for the assessment of colonic diseases, both inflammatory and neoplastic, on a more "functional" level. DW-MR images may be evaluated both qualitatively and quantitatively: the former results from a visual assessment of the DW-MR sequences, in which areas of restricted diffusion will appear hyperintense against a hypointense background on highest b-values obtained and hypointense on corresponding ADC map; quantitative evaluation of the water diffusion characteristics is performed by expressing them as an apparent diffusion coefficient (ADC) value. Both qualitative and quantitative information have been the subject of several works on colonic diseases.

A critical review of colonic polyp and cancer detection, characterization of colonic wall thickening, and assessment of inflammatory bowel disease of the colon will be performed in this chapter. A perspective on future applications and trends will also be discussed.

L. C. Semedo
Medical Imaging Department, Coimbra Hospital and University Centre, Faculty of Medicine, University of Coimbra, Coimbra, Portugal

© Springer International Publishing AG, part of Springer Nature 2019
S. Gourtsoyianni, N. Papanikolaou (eds.), *Diffusion Weighted Imaging of the Gastrointestinal Tract*, https://doi.org/10.1007/978-3-319-92819-7_4

53

4.2   Technical Considerations

Imaging of the large bowel with DWI is challenging. In fact, relatively long acquisition times causing an increased sensitivity to bowel motion and the presence of T2 shine-through effect, which is frequently encountered in the bowel lumen, may hamper the diagnostic information conveyed by MR images [1].

A combination of high magnetic field MR scanners, decreased acquisition times, multichannel coils, and parallel imaging techniques have been helpful to reduce these technical limitations. New pulse sequences such as echo-planar imaging have contributed to a reduction in acquisition time, therefore also decreasing sensitivity to bowel and respiratory movements [1]. At present, no clear benefit from the use of antiperistaltic agents has been clearly shown, although theoretically they may reduce bowel motion artifacts.

The T2 shine-through effect may be reduced with the use of high b-values and short echo times.

For imaging of the bowel, both a low b-value ranging from 0 to 50 s/mm^2 and at least one high b-value (800–1000 s/mm^2) are useful. In our daily practice, usually three b-values ($b = 50$, $b = 500$, and $b = 1000$ s/mm^2) are routinely acquired in a high-field magnet. This allows for the calculation of ADC values minimizing the effects of microperfusion allowing at the same time the acquisition of images with high contrast and signal-to-noise ratios.

Axial images should be preferably acquired, since they are less prone to motion artifacts than those acquired in other planes, namely in the coronal plane [1].

For image acquisition, different approaches may be used: e.g., a navigator-triggered technique, which helps reduce motion artifact and increases signal-to-noise ratio, the main drawback being the increased acquisition time. Sequences acquired with the patient breathing freely or in several breath holds are generally faster but suffer from motion artifacts and less detail [1].

The use of oral and rectal cleansing before DW-MRI is questionable nowadays. If it seems reasonable to think that colonic distention would improve detection and evaluation of lesions, it is also true that there is a growing body of evidence showing that colonic assessment without bowel preparation is feasible, yields satisfactory results, and improves patient compliance by reducing discomfort.

4.3   Detection of Polyps and Cancer

The rationale behind the use of DW-MRI for the detection of colonic polypoid lesions and cancer is that those lesions will exhibit restriction to diffusion and therefore will show high signal intensity at high b-value DW images. As such, the high lesion-to-background contrast that can be provided by DW-MRI will theoretically

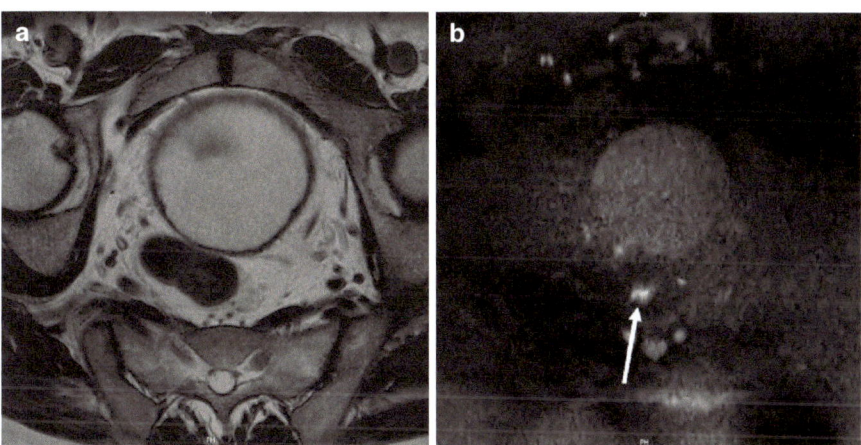

Fig. 4.1 64-year-old male with a history of occult blood loss in stool. On the T2-weighted image (**a**), no abnormalities are found, whereas on the DWI sequence (b1000), a small focus of high signal intensity is apparent on the topography of the upper rectum (white arrow on **b**), which corresponded to a sessile polyp on colonoscopy. Both sequences were obtained without previous bowel preparation

provide an advantage compared to conventional sequences (Fig. 4.1). Furthermore, DWI does not require the administration of any contrast agents and can be performed without any bowel preparation [2].

A feasibility study published by Dutch authors on the detection of polyps included 26 patients and achieved a lesion-based sensitivity of 80.0% for clinically relevant lesions (polyps ≥6 mm and cancer) [2]. Nevertheless, the authors suggested that further technical developments are required in order to increase the diagnostic yield of DW-MRI in the detection of polypoid lesions of the colon.

Some authors have also investigated the value of DW-MRI in the detection of colorectal cancer. Ichikawa et al. retrospectively assessed the diagnostic value of DWI for the detection of colorectal cancer in 33 patients with neoplasms and 15 controls and reported a sensitivity of 90.9% and specificity of 100% for the diagnosis of colonic adenocarcinoma [3]. Neoplastic lesions, due to its high cellularity, presented with high signal intensity on the high *b*-value images, by contrast to a dark background (Fig. 4.2). On the ADC maps, these lesions characteristically appeared with low signal intensity. Again, the detection of neoplastic lesions could be achieved on non-gadolinium-enhanced sequences, and previous bowel cleaning was not performed.

In polypoid cancers, DWI can clearly demarcate the areas within the polyp that show a high cellular content, helping to distinguish them from the low-cellular areas (Fig. 4.3).

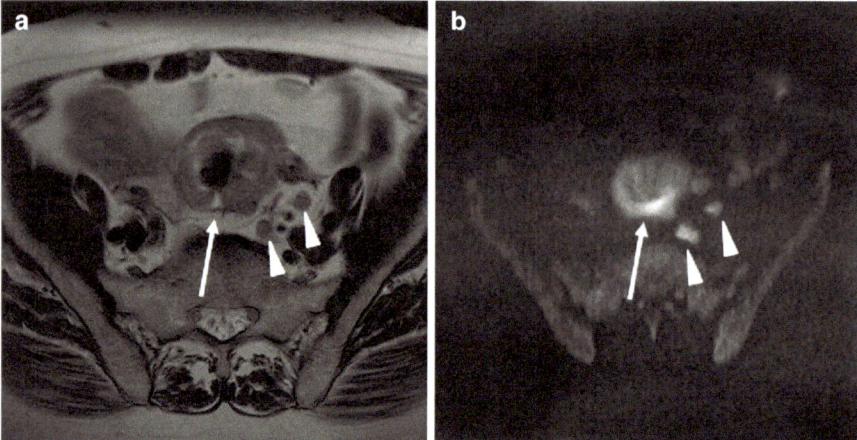

Fig. 4.2 77-year-old female with an endoluminal lesion on colonoscopy. On the T2-weighted image, an irregular wall thickening is disclosed at the level of the distal sigmoid (white arrow on **a**) with some perilesional adenopathies (white arrowheads on **a**). On the DWI sequence (b1000), both the primary lesion (white arrow on **b**) and the adenopathies (white arrowheads on **b**) reveal predominant hyperintensity against a dark background

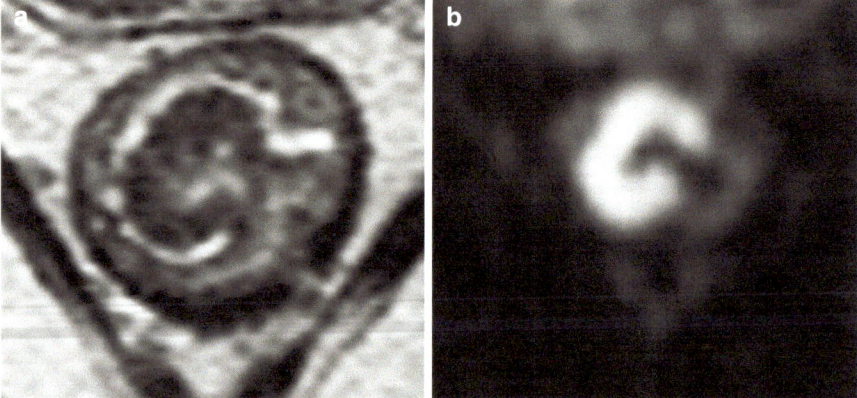

Fig. 4.3 62-year-old male with a polyp on colonoscopy. On the T2-weighted image, a large peduncleulated polypoid cancer is seen at the level of the rectum (**a**). On the DWI sequence (b1000), the head of the polyp demonstrates high signal intensity, corresponding to the more cellular areas, in contrast to the stalk, which, because of the relatively low cellularity, appears dark on DWI (**b**)

4.4 Characterization of Wall Thickening

DW-MRI has been used as a tool to help characterize diffuse bowel wall thickening, namely to distinguish between malignancy and various benign conditions, including inflammatory, ischemic, or infectious bowel diseases.

Solak et al. retrospectively evaluated 26 patients with malignant disease and 15 patients with benign conditions of the colorectum by DW-MRI, visually assessing high b-value ($b = 800$ s/mm^2) DWI images and ADC maps, and also quantifying ADC values, having defined endoscopic biopsy as the gold standard [4]. Results from this study demonstrated that the difference between the mean ADC values of benign and malignant conditions was statistically significant, with ADC values of benign lesions being significantly higher than those of malignant lesions. By applying a cutoff value of 1.21×10^{-3} mm^2/s, ADC yielded a sensitivity of 100%, a specificity of 87.3%, and an accuracy of 89.3% in the discrimination of malignant colorectal pathology. With the combined visual assessment of the high b-value images and the measurement of ADC values, malignant and benign lesions could be differentiated with 100% sensitivity, 89.2% specificity, and 90.4% accuracy. Importantly also, although some benign lesions were interpreted as malignant, no malignant lesion was judged to be benign on the visual assessment [4].

Other authors directed their attention to the differentiation between a particular inflammatory condition (acute diverticulitis) and cancer. Both clinical conditions may show overlapping signs, symptoms, and imaging features—particularly on CT—and the coexistence of acute diverticulitis superimposed on a colon cancer may obscure the latter on imaging. Therefore, DW-MRI was tested as an alternative to CT in establishing the diagnosis of diverticulitis [5]. Öistämö et al. retrospectively examined patients presenting with either diverticulitis or sigmoid cancer with DW-MRI. This study reported a sensitivity and specificity for the diagnosis of colon cancer and diverticulitis of 100% when using DW-MRI, whereas the sensitivity and specificity for the diagnosis of colon cancer and diverticulitis were 67% and 93%, respectively, using CT [5]. However this study included only two groups of 15 patients, and its results should be confirmed and further validated in larger studies.

In these works, the differentiation between malignant and benign lesions of the colonic wall relied on the identification of hyperintense areas within the colonic wall on high b-value images and lower ADC values, in the former group of diseases.

4.5   Assessment of Inflammatory Bowel Disease

One major application of DW-MRI consists of the evaluation of inflammatory bowel diseases. In fact, the largest body of evidence on the role of DW-MRI in the colon comes from the assessment of ulcerative colitis and Crohn's disease. Several works have been published in the field, regarding detection of colonic inflammation, assessment of disease activity, and evaluation of response to therapy.

4.5.1   Detection of Inflammatory Changes in the Colon

A feasibility study by Oto et al. was designed to determine the possibility of a role for DWI in the detection of bowel inflammation and to investigate the changes in ADC values of the inflamed bowel in patients with Crohn's disease, with pathologic

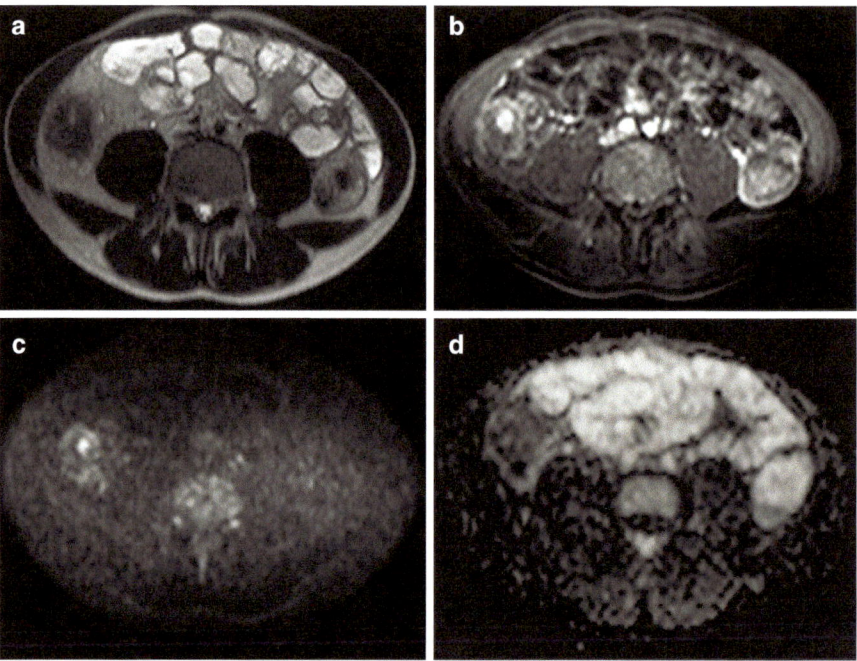

Fig. 4.4 31-year-old male with Crohn's disease. On the T2-weighted image, there is thickening of the wall of the right colon (**a**) with correspondent hyperenhancement on the fat-suppressed, gadolinium-enhanced, T1-weighted image (**b**). On the DWI sequence (b1000), there is restriction to diffusion with hyperintensity of the colonic wall (**c**) and low signal intensity on the ADC map (**d**), compatible with inflammatory changes at that level

features as the gold standard [6]. Inflammation of the bowel wall causes restricted diffusion, and as such DWI yields both qualitative (increased signal intensity) and quantitative (decreased ADC values) information that can be helpful in the evaluation of bowel inflammation (Fig. 4.4). In addition to the increased number of inflammatory cells, dilated lymphatic channels, hypertrophied neuronal tissue, and the development of granulomas in the bowel wall can further narrow the extracellular space and therefore contribute to the restricted diffusion of water molecules.

Diffusion-weighted magnetic resonance colonography (DW-MRC) without oral or rectal preparation proved to be a reliable tool for detecting colonic inflammation in several studies. The technique does not need fasting, is noninvasive, and does not require any bowel preparation. Oussalah et al. studied 96 patients with both ulcerative colitis and Crohn's disease (68 had concomitant endoscopy) with DW-MRC without fasting or oral or rectal preparation [7]. On DW-MRC, six radiological signs were studied: (1) DWI hyperintensity, (2) rapid gadolinium enhancement after intravenous contrast medium administration, (3) differentiation between the mucosa-submucosa complex and the muscularis propria, (4) bowel wall thickening, (5) parietal edema, and (6) the presence of ulceration(s). In the ulcerative colitis group, the presence of a DWI hyperintensity demonstrated a sensitivity and a specificity of 90.79% and 80%, respectively, for the detection of endoscopic inflammation, with an area under the ROC curve of 0.854. DWI hyperintensity was statistically more effective for the detection of endoscopic colonic inflammation in ulcerative colitis than in Crohn's disease. Comparatively, in the ulcerative colitis group, rapid gadolinium enhancement correlated with endoscopic inflammation in the colon with a sensitivity and specificity of 72.37% and 96.67%, respectively, with an area under the ROC curve of 0.845. Rapid gadolinium enhancement was significantly more effective for the detection of endoscopic inflammation in ulcerative colitis than in Crohn's disease. Of note, there was no statistically significant difference in accuracy between DWI hyperintensity and rapid gadolinium enhancement areas under the ROC curves in ulcerative colitis and Crohn's disease. In ulcerative colitis, ROC analyses for the four remaining parameters of the MR score for the detection of endoscopic inflammation demonstrated a good sensitivity (88.16%) and specificity (83.33%) for the "differentiation between the mucosa-submucosa complex and the muscularis propria." For the three other items, the sensitivity was low, ranging from 38.16% to 67.11%, with excellent specificities ranging from 93.33% to 96.67%. In Crohn's disease, ROC analyses for the same four parameters revealed low sensitivities ranging from 36.11% to 62.5% and good to excellent specificities ranging from 75% to 100%. Among these four parameters, the "differentiation between the mucosa-submucosa complex and the muscularis propria" and "ulcerations" exhibited better accuracy for the detection of endoscopic inflammation in ulcerative colitis than in Crohn's disease. The accuracy was similar for the other two items in both ulcerative colitis and Crohn's disease. Logistic regression analysis showed that DWI hyperintensity was predictive of the presence of endoscopic inflammation in both the ulcerative colitis and the Crohn's disease groups (odds ratio = 13.26 and 2.67, respectively) [7].

Similarly, Sirin et al. used DW-MRC to assess whether intravenous contrast was needed to depict inflammatory lesions in the bowel when DWI was also available, in a pediatric population [8]. In this retrospective study, patients received bowel preparation, and optical colonoscopy was the gold standard for the 37 individuals studied. Mean sensitivity and specificity for two readers for the depiction of inflammatory lesions were, respectively, 78.4% and 100% using gadolinium-enhanced T1-weighted MRC, 95.2% and 100% using DWI, and 93.5% and 100% combining both imaging techniques compared with colonoscopy (including results of the histopathological samples). In six patients, inflammatory lesions were only detected by DWI; in another six patients, DWI detected additional lesions. The preferred b-value with the best detectability of the lesions was $b = 1000$ s/mm^2 in 28 of the 30 patients (93.3%) (8).

A recent study also assessed the role of DW-MRI without bowel preparation in the detection of ulcerative colitis in 20 patients with optical colonoscopy as the gold standard [9]. The authors assessed the following imaging signs: (1) DWI hyperintensity ($b = 800$ s/mm^2), (2) rapid gadolinium enhancement after intravenous contrast medium administration (20–25 s after gadolinium infusion), (3) differentiation between the mucosa-submucosa complex and the muscularis, (4) bowel wall thickening (exceeding 5 mm), (5) parietal edema, (6) the presence of ulceration(s), and (7) comb sign of engorged vasa recta that perpendicularly penetrated the bowel wall. The results showed that DWI provided qualitative and quantitative information when this technique was combined with conventional magnetic resonance imaging without bowel preparation; the combined technique demonstrated a good diagnostic performance to detect colonic inflammation in ulcerative colitis. DWI hyperintensity at $b = 800$ s/mm^2 detected endoscopic colonic inflammation with a sensitivity of 93.0% and a specificity of 79.3% with an area under the ROC curve of 0.867. With rapid gadolinium enhancement, endoscopic colonic inflammation was detected with a sensitivity of 73.2%, a specificity of 93.1%, and an area under the ROC curve of 0.853. The accuracy between DWI hyperintensity and rapid gadolinium enhancement was not significantly different. Differentiation between the mucosa-submucosa complex and the muscles revealed a good sensitivity (80.3%) and specificity (86.2%). The four other signs demonstrated low sensitivities (range, 43.7–66.2%) and excellent specificities (range, 89.7–93.1%) [9].

From the analysis of the published works, it seems reasonable to recognize that DW-MRC, which combines morphological MRI and DWI, even without oral or rectal preparation might be used in clinical practice to evaluate colonic inflammation, particularly in ulcerative colitis. DWI has the potential to replace gadolinium-enhanced sequences in order to detect inflammatory changes in the colon, therefore reducing the likelihood of adverse effects from the use of gadolinium-based contrast media and, at the same time, reducing examination costs.

4.5.2 Assessment of Disease Activity

With the aim of assessing disease activity in patients with ulcerative colitis, Kılıçkesmez et al. prospectively studied 28 patients in different stages of the disease by means of DW-MRI without preparation, measuring ADC of the bowel wall in sigmoid colon and rectum and comparing the findings with endoscopy [10]. Results disclosed no statistically significant difference in the ADC of the sigmoid colon in patients with active, subacute, and remissive ulcerative colitis. On the contrary, ADC values of the rectum were statistically different between patients in the active ($1.08 \pm 0.14 \times 10^{-3}$ mm^2/s) and subacute phases ($1.13 \pm 0.23 \times 10^{-3}$ mm^2/s) of disease and those in remission ($1.29 \pm 0.17 \times 10^{-3}$ mm^2/s). Therefore, an increased activity of the disease was correlated with lower ADC values [10].

Similarly, Kiryu et al. investigated the application of free-breathing DW-MRI to assess active Crohn's disease [11]. The findings of a conventional barium study or surgery were regarded as the gold standard. The ADC was significantly lower in the disease-active segments than in the disease-inactive segments in the large bowel ($1.52 \pm 0.43 \times 10^{-3}$ mm^2/s versus $2.31 \pm 0.59 \times 10^{-3}$ mm^2/s, respectively). The sensitivity, specificity, and accuracy were 85.7%, 75.7%, and 77.3%, respectively, in the large bowel. The accuracy was 82.6% in the ascending colon, 85.0% in the transverse colon, 80.8% in the descending colon, 72.4% in the sigmoid colon, and 70.0% in the rectum [11].

Diffusion-weighted magnetic resonance enterocolonography (DW-MREC) with no bowel cleansing and no rectal enema was performed by Buisson et al. to prospectively evaluate patients with Crohn's disease, specifically for the indirect detection of ulcerations in this setting [12]. Forty-four patients were studied and results were compared to the gold standard (ileocolonoscopy): a total of 158 colorectal segments were assessed. The authors showed that not only the segmental ADC measured on the bowel wall in these segments was correlated with endoscopic scores but also that MRI accuracy to detect endoscopic ulcerations, using a ADC < 1.88 for colon/

rectum, ranged from 63.2% (cecum/right colon) to 84.6% (left/sigmoid colon). This work also disclosed a relationship between ulcer size and ADC: the segmental ADC values decreased when the ulceration size increased [12].

Sato et al. also aimed to compare the findings of DW-MREC with endoscopically identified lesions according to inflammatory grades and assess the diagnostic accuracy of DW-MREC for sensitivity and grading severity in Crohn's disease [13]. A total of 27 patients were evaluated. A positive lesion was defined as having at least one of the following: wall thickness, edema, high intensity on DWI images, and relative contrast enhancement on MREC. The sensitivities were 100% for ulcer, 84.6% for erosion, and 52.9% for redness, suggesting an ability to detect milder lesions such as erosion or redness in MREC. For DWI in specific, the sensitivities for endoscopic ulcer, erosion, and redness were 80.9%, 69.2%, and 33.3%, respectively. The specificities for endoscopically identified lesions were high (92.1%). When the lesion was defined as having two or three positive MREC findings, sensitivity increased: sensitivities for either wall thickness or DWI high intensity, either edema or DWI high intensity, and either wall thickness or edema or DWI high intensity were 80.9%, 80.9%, and 80.9% for ulcer, respectively; 76.9%, 69.2%, and 76.9% for erosion, respectively; and 41.2%, 35.3%, and 41.2% for redness, respectively. Specificities were 92%, 92%, and 92%, respectively [13].

Therefore, DWI could be an adjunct to differentiate between active inflammation and quiescent disease (Fig. 4.5).

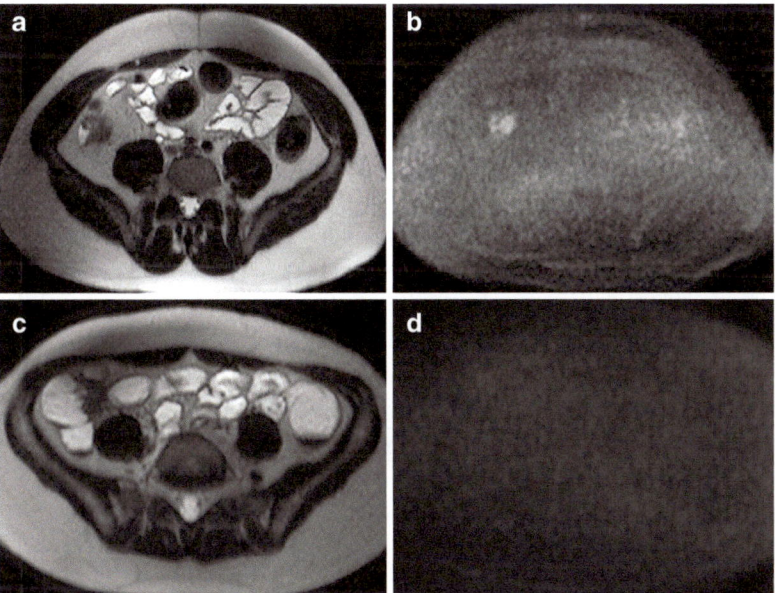

Fig. 4.5 Two different patients (**a, b**, 19-year-old male; **c, d**, 16-year-old female) with Crohn's disease showing areas of wall thickening at the level of the right colon on the T2-weighted images (**a, c**). On the DWI sequence (b1000), there is restriction of diffusion with hyperintensity of the colonic wall in the first patient (**b**) indicating active disease, whereas in the second patient, no areas of restricted diffusion are seen, corresponding to quiescent disease

4.5.3 Evaluation of Response to Therapy

A recently published paper aimed to assess DW-MREC parameters as predictors of remission after anti-TNF induction therapy in Crohn's disease [14]. Forty consecutive patients were enrolled in this prospective study, being evaluated by DW-MREC with no rectal distension and no bowel cleansing. Patients were evaluated before treatment and at week 12. The authors showed that a mean ADC cutoff of 1.96×10^{-3} mm^2/s was predictive of remission at week 12 (area under the ROC curve = 0.703) with sensitivity, specificity, positive predictive value, and negative predictive value of 70.0%, 65.0%, 66.7%, and 68.4%, respectively. In a multivariate analysis, mean ADC $< 1.96 \times 10^{-3}$ mm^2/s (odds ratio = 4.87), reflecting high inflammatory activity, was predictive of remission at week 12. These results suggest that DW-MREC may help to select patients with objective digestive inflammation who could benefit from anti-TNF therapy and could be helpful to predict remission after anti-TNF induction therapy [14]. However, results from this pilot study need to be confirmed in an independent larger cohort.

Similarly, Sakuraba et al. evaluated 13 individuals 1 year after infliximab induction therapy by DW-MRI scans which were assessed as predictors of maintained response, or remission, through 3 years of treatment in patients with CD [15]. Examinations were performed 1 and 3 years after the starting point of the infliximab therapy. DW-MRI predicted the presence of synergistic mucosal changes on colonoscopy with a sensitivity of 80.52% and specificity of 66.67%. DW-MRI at 1 year was able to predict the presence of endoscopic inflammation with a sensitivity of 66.67%, a specificity of 80.52%, and an area under the ROC curve of 0.7359. Also, DW-MRI at 3 years suggested endoscopic inflammation with a sensitivity of 94.12%, a specificity of 73.91%, and an area under the ROC curve of 0.8402 [15].

Evaluation of therapy response by DW-MRI is naturally regarded with increasing interest by both clinicians and radiologists working on the field of inflammatory bowel disease, since the patient's discomfort and risk of injury are minimized because of the noninvasiveness of the method. Furthermore, lesions that are not accessible by endoscopy because of stenosis or adhesion can be evaluated, as well as the extraintestinal tissues. In the future, biomarkers of response to treatment based on DW-MRI might be helpful for optimizing the indications for endoscopy and further treatment of these patients.

4.6 Future Applications and Perspectives

Many expectations are being raised by the application of DW-MRI in the field of colorectal oncology. DW-MRI may theoretically be used to assess and monitor therapy response of colorectal cancer.

An animal study by Schneider et al. performed the early monitoring of antiangiogenic therapy in an experimental tumor model [16]. Using quantitative DW-MRI, the authors found that therapy of human colon carcinoma xenografts with the multi-tyrosine kinase inhibitor regorafenib significantly increased water diffusivity in tumorous tissue after 6 days of treatment. Regorafenib significantly reduced tumor growth compared to the control group. Using either tumor ADC changes or tumor growth to distinguish between therapy and control group resulted in a diagnostic accuracy of about 78% and 83%, respectively, which was improved by the approach to combine both parameters using Fisher's linear discriminant analysis to about 96%, thus highlighting the potential of multiparameter MRI as an imaging biomarker for noninvasive monitoring of early tumor therapy and allowing in this way a more patient-tailored therapeutic approach [16].

Another experimental study aimed to assess the potential value of combined MR elastography and DW-MRI in the detection of microstructural changes of murine colon tumors during growth and antivascular treatment for two models of implantation (ectopic and orthotopic) [17]. DW-MRI was sensitive to tumor cell alterations, including cellularity and micronecrosis; ADC decreased significantly for the ectopic model between early and angiogenic stages, whereas no significant ADC change was observed for the orthotopic model between these stages. MR elastography allowed monitoring of changes in vascularization. The authors concluded that MR elastography and DW-MRI have the potential of being complementary for noninvasive surveillance of tumor evolution [17].

In addition to conventional ADC measurements in the monoexponential range, the development of high-performance gradient coils enables DWI measurements with stronger diffusion weighting using higher b-values (e.g., 1500 s/mm^2) and increased diffusion contrast. Under these conditions the signal attenuation is often non-monoexponential. This is a consequence of restricted diffusion as the mean-squared displacements of diffusing protons are no longer Gaussian distributed. Quantitative non-Gaussian diffusion models have been developed to fit diffusion signals with high b-values. Several of these non-Gaussian diffusion models have been implemented in cancer imaging and appear to show new information or higher sensitivity compared with conventional ADC measurements [18].

In fact, Xu et al. tested both conventional ADC and non-Gaussian model measurements and analyses in order to assess the early therapeutic response of human colon cancer to barasertib [19]. The results suggest that the non-Gaussian DWI model-derived parameters were capable of detecting earlier tumor changes to treatment in comparison with conventional ADC. Non-Gaussian DWI may potentially provide an opportunity to better evaluate tumor status earlier than ADC and tumor volume changes that are currently widely used in clinical cancer research, therefore yielding an opportunity to assist clinicians to better enable necessary therapeutic adjustments in a timely manner to enhance treatment efficacy and avoid unnecessary treatment delays, toxicity, and expenses [19].

The role of DWI-MR as a biomarker of response in colon cancer should be, at present, regarded as an adjunct to clinical tools (e.g., endoscopy and biopsy). The results published so far are obviously still premature for clinical decision-making, but their promise warrants further validation by large and prospective patient studies.

References

1. Dohan A, Taylor S, Hoeffel C, et al. Diffusion-weighted MRI in Crohn's disease: current status and recommendations. J Magn Reson Imaging. 2016;44(6):1381–96.
2. Leufkens AM, Kwee TC, van den Bosch MA, Mali WP, Takahara T, Siersema PD. Diffusion-weighted MRI for the detection of colorectal polyps: feasibility study. Magn Reson Imaging. 2013;31(1):28–35.
3. Ichikawa T, Erturk SM, Motosugi U, et al. High-B-value diffusion-weighted MRI in colorectal cancer. AJR Am J Roentgenol. 2006;187:181–4.
4. Solak A, Genç B, Solak I, et al. The value of diffusion-weighted magnetic resonance imaging in the differential diagnosis in diffuse bowel wall thickening. Turk J Gastroenterol. 2013;24(2):154–60.
5. Öistämö E, Hjern F, Blomqvist L, Von Heijne A, Abraham-Nordling M. Cancer and diverticulitis of the sigmoid colon. Differentiation with computed tomography versus magnetic resonance imaging: preliminary experiences. Acta Radiol. 2013;54(3):237–41.
6. Oto A, Zhu F, Kulkarni K, Karczmar GS, Turner JR, Rubin D. Evaluation of diffusion-weighted MR imaging for detection of bowel inflammation in patients with Crohn's disease. Acad Radiol. 2009;16(5):597–603.
7. Oussalah A, Laurent V, Bruot O, et al. Diffusion-weighted magnetic resonance without bowel preparation for detecting colonic inflammation in inflammatory bowel disease. Gut. 2010;59(8):1056–65.
8. Sirin S, Kathemann S, Schweiger B, et al. Magnetic resonance colonography including diffusion-weighted imaging in children and adolescents with inflammatory bowel disease: do we really need intravenous contrast? Invest Radiol. 2015;50(1):32–9.
9. Yu LL, Yang HS, Zhang BT, et al. Diffusion-weighted magnetic resonance imaging without bowel preparation for detection of ulcerative colitis. World J Gastroenterol. 2015;21(33):9785–92.
10. Kilickesmez O, Atilla S, Soylu A, et al. Diffusion-weighted imaging of the rectosigmoid colon: preliminary findings. J Comput Assist Tomogr. 2009;33(6):863–6.
11. Kiryu S, Dodanuki K, Takao H, et al. Free-breathing diffusion-weighted imaging for the assessment of inflammatory activity in Crohn's disease. J Magn Reson Imaging. 2009;29(4):880–6.
12. Buisson A, Hordonneau C, Goutte M, Boyer L, Pereira B, Bommelaer G. Diffusion-weighted magnetic resonance imaging is effective to detect ileocolonic ulcerations in Crohn's disease. Aliment Pharmacol Ther. 2015;42(4):452–60.
13. Sato H, Tamura C, Narimatsu K, et al. Magnetic resonance enterocolonography in detecting erosion and redness in intestinal mucosa of patients with Crohn's disease. J Gastroenterol Hepatol. 2015;30(4):667–73.
14. Buisson A, Hordonneau C, Goutte M, et al. Diffusion-weighted magnetic resonance enterocolonography in predicting remission after anti-TNF induction therapy in Crohn's disease. Dig Liver Dis. 2016;48(3):260–6.
15. Sakuraba H, Ishiguro Y, Hasui K, et al. Prediction of maintained mucosal healing in patients with Crohn's disease under treatment with infliximab using diffusion-weighted magnetic resonance imaging. Digestion. 2014;89(1):49–54.
16. Schneider MJ, Cyran CC, Nikolaou K, Hirner H, Reiser MF, Dietrich O. Monitoring early response to anti-angiogenic therapy: diffusion-weighted magnetic resonance imaging and volume measurements in colon carcinoma xenografts. PLoS One. 2014;9(9):e106970.
17. Jugé L, Doan BT, Seguin J, et al. Colon tumor growth and antivascular treatment in mice: complementary assessment with MR elastography and diffusion-weighted MR imaging. Radiology. 2012;264(2):436–44.
18. Maier SE, Sun Y, Mulkern RV. Diffusion imaging of brain tumors. NMR Biomed. 2010;23:849–64.
19. Xu J, Li K, Adam Smith R, et al. A comparative assessment of preclinical chemotherapeutic response of tumors using quantitative non-Gaussian diffusion MRI. Magn Reson Imaging. 2017;37:195–202.

Rectum

5

Doenja M. J. Lambregts and Regina G. H. Beets-Tan

Abbreviations

ADC   Apparent diffusion coefficient
AUC   Area under the (ROC) curve
CRT   Chemoradiotherapy
DWI   Diffusion-weighted (magnetic resonance) imaging
EPI   Echo planar imaging
MRF   Mesorectal fascia
MRI   Magnetic resonance imaging
N-stage   Nodal stage
T-stage   Tumour stage
TME   Total mesorectal excision

5.1   Introduction

In the last decade, well over 100 papers have been published on the use of DWI for rectal cancer imaging. In an increasing number of centres worldwide, a DWI sequence is now routinely included in the rectal MRI protocol, not only for research purposes but also for clinical rectal cancer assessment. The routine use of DWI is also recommended by the expert consensus guidelines from the European Society of Gastrointestinal and Abdominal Radiology (ESGAR), particularly for restaging of rectal tumours after chemoradiotherapy [1]. This chapter will discuss various applications of DWI for rectal cancer imaging, in the

D. M. J. Lambregts (✉) · R. G. H. Beets-Tan
Department of Radiology, The Netherlands Cancer Institute, Amsterdam, Netherlands
e-mail: d.lambregts@nki.nl; r.beetstan@nki.nl

© Springer International Publishing AG, part of Springer Nature 2019
S. Gourtsoyianni, N. Papanikolaou (eds.), *Diffusion Weighted Imaging of the Gastrointestinal Tract*, https://doi.org/10.1007/978-3-319-92819-7_5

primary staging setting, as well as for restaging of tumours after neoadjuvant treatment. Finally, the use of DWI for follow-up after treatment, its potential value as a prognostic marker and main interpretation pitfalls for rectal DWI will be discussed.

5.2 DWI for Primary Rectal Cancer Staging

5.2.1 DWI for Rectal Tumour Detection

Only a few reports have focussed on the primary detection of rectal tumours using DWI. This makes sense, since the primary detection of rectal tumours is not so much a task for the radiologist. Once a patient is referred for imaging, the diagnosis is typically already clear from clinical examination and endoscopy with biopsy. The main role of imaging is therefore not to detect, but rather to stage the tumour. The studies that did focus on rectal tumour detection reported very high detection rates for DWI between 93% and 100% [2, 3]. DWI may be helpful to improve tumour conspicuity in specific cases, particularly for less experienced readers, for example, when the tumour is small and/or when there is a lot of faeces present in the rectal lumen, as a result of which the tumour may be more difficult to discern (Fig. 5.1). DWI is less useful for the assessment of mucinous-type rectal adenocarcinoma since—as opposed to tubular-type adeno-carcinomas that have a dense cellular structure leading to diffusion restriction with high signal on DWI—mucinous tumours typically do not show restricted diffusion but have a relatively low signal intensity on high b-value DWI with corresponding high ADC signal (Fig. 5.2) [4].

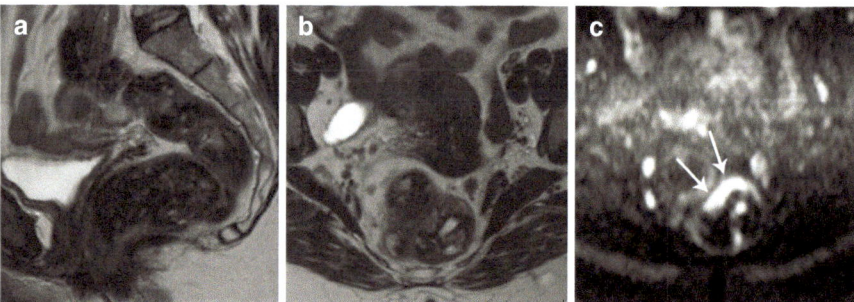

Fig. 5.1 Example of a female patient with a small tumour in the mid-rectum. On the sagittal (**a**) and transverse (**b**) T2-weighted images, the tumour is hardly visible and difficult to discern within the stool present in the rectal lumen. On b1000 DWI (**c**), the small tumour is easily recognised (arrows)

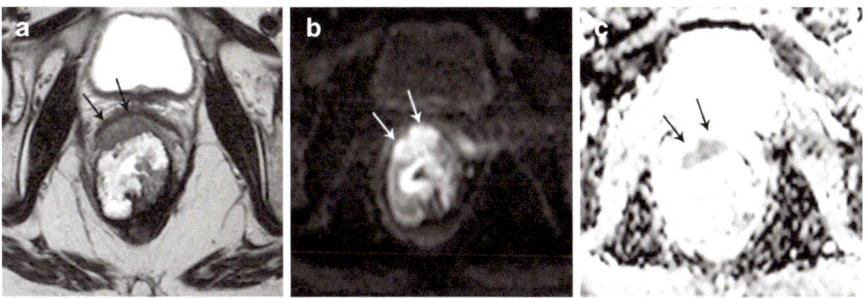

Fig. 5.2 Transverse T2-weighted image (**a**), b1000 DWI (**b**) and ADC map (**c**) of a female patient with a largely mucinous tumour in the distal rectum. The mucinous components of the tumour show high signal on T2-weighted MRI with corresponding high signal on the ADC map and only slightly increased signal on b1000 DWI (caused by T2 shine through effects). On the anterior side, the tumour has a small solid component (arrows) with restricted diffusion resulting in high signal on b1000 DWI and corresponding low signal in the ADC map

5.2.2 DWI for Rectal Tumour Staging

Key principles of rectal tumour staging are to determine the depth of tumour invasion into and beyond the different layers of the bowel wall and the minimum distance between the tumour and the mesorectal fascia (MRF) and to assess whether the tumour invades any nearby organs or structures. These factors are mainly assessed using T2-weighted MRI sequences in multiple planes parallel and perpendicular to the tumour axis as these provide good anatomical detail to visualise the morphology of the rectal wall, the mesorectal compartment and the mesorectal fascia. There appears to be little role for DWI in primary tumour staging. The few studies that investigated the use of DWI for local tumour staging reported similar sensitivities ranging between 64% and 90% for T1-T2 tumours and 88% and 95% for T3-T4 tumours and specificities of 88–95% for T1-T2 tumours and 82–100% for T3-T4 tumours, results similar to those found for T2-weighted MRI with no significant contributing value for DWI [5, 6]. The value of DWI to assess mesorectal fascia involvement has so far only been reported in the restaging setting (see Sect. 5.3.2).

5.2.3 DWI for Lymph Node Staging

Staging of rectal cancer lymph nodes remains one the most difficult tasks for the radiologist. Theoretically, DWI could be a suitable technique to assess lymph nodes, since lymphoid tissue has a dense cellular structure which makes it easily detectable on DWI (Fig. 5.3). Different reports have indeed shown that DWI is a highly sensitive technique to detect rectal cancer lymph nodes and that it can help identify over

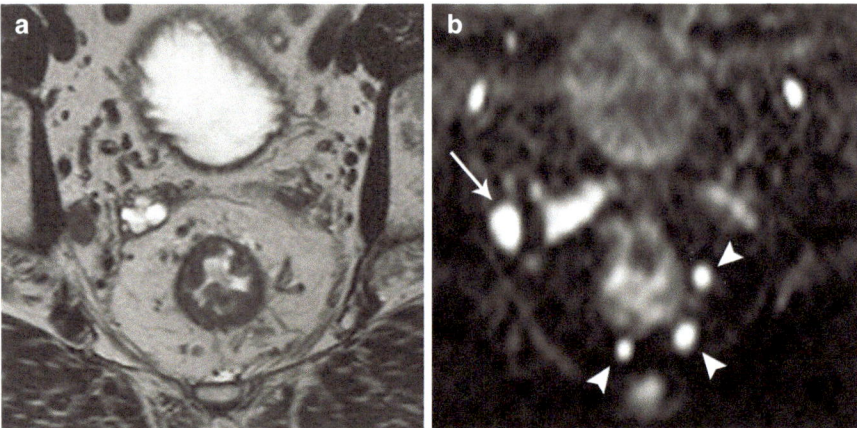

Fig. 5.3 T2-weighted (**a**) and b1000 diffusion-weighted (**b**) images that illustrate how DWI can be used to help detect lymph nodes. The mesorectal lymph nodes (arrowheads) show markedly bright signal on DWI making them easier to detect than on the corresponding T2-weighted image. Also note the enlarged extramesorectal lymph node in the right obturator area (arrow)

25% more nodes compared to routine morphological T2-weighted MR sequences [7–11]. There is however no consensus on the value of DWI to discriminate between benign/reactive and metastatic lymph nodes. Heijnen et al. reported that benign and metastatic rectal cancer lymph nodes show equally high signal intensities on DWI. When using the signal intensity of nodes on DWI as a criterion to discriminate malignant nodes, results were disappointing with areas under the receiver-operator characteristics (ROC) curve (AUC) of only 0.45–50 [7]. Similarly, in a study by Mizukami and colleagues, positive predictive value was only 52% when using high signal on DWI as a criterion for malignancy, indicating that almost half of the nodes that show high signal on DWI are in fact benign [8]. Two studies quantitatively studied nodal ADC values to differentiate between benign and metastatic nodes at the time of primary staging. Significant differences in nodal ADC between benign and metastatic nodes were found, but sensitivity (67–78%) and specificity (60–67%) for nodal ADC were suboptimal [7, 10] and do not offer a considerable improvement compared to the use of the more widely used MRI criterion nodal size, with reported sensitivities and specificities of 55–78% [12, 13]. Moreover, in many of the typically small rectal cancer nodes, it can be very difficult to reproducibly measure ADC [11]. Altogether, the main benefit of DWI for primary nodal staging thus appears to help detect rather than help characterise the nodes. One report showed that studying morphologic characteristics of nodes on DWI may be helpful to characterise them and that metastatic nodes typically show a more heterogeneous signal and ill-defined border on DWI [14]. Similar results have however previously also been reported for the assessment of nodal morphology using routine T2-weighted sequences [15, 16].

5.3   DWI for Tumour Restaging After Chemoradiotherapy

5.3.1   DWI for Tumour Response Assessment

Interestingly, the vast majority of papers that were published in recent years on DWI in rectal cancer focussed on its use for assessment of tumour response to neoadjuvant chemoradiation treatment. This striking focus on response evaluation can probably be explained by the fact that we are currently witnessing a paradigm shift towards less invasive treatments in rectal cancer patients who show a very good response to neoadjuvant chemoradiotherapy. Patients with only a small tumour remnant after CRT may be treated with a local excision of the tumour remnant instead of total mesorectal excision [17]. Moreover, trials from Brazil, Denmark, the Netherlands and the USA have shown that in patients with a complete tumour regression after CRT, a nonoperative management with stringent follow-up ('wait-and-see') may be considered as an alternative to surgery with comparable survival outcomes [18–21]. This shift in treatment makes response evaluation an increasingly relevant issue. Morphological MRI has limited accuracy for re-evaluation of the tumour stage after neoadjuvant treatment. In a meta-analysis, reported pooled specificity for T-staging after CRT was 91%, but sensitivity was only 50%. Particularly the sensitivity to differentiate between a complete tumour response and residual tumour was very low (19%) [22]. The main problem is the evaluation of fibrosis after CRT. As a result of radiotherapy, tumours undergo fibrotic changes which appear hypointense on T2-weighted MRI. In these hypointense areas, it can be very difficult to discern small areas of viable residual tumour. There is increasing evidence that DWI can help in these cases and aid to improve the performance of MRI for tumour restaging after CRT.

There are several ways DWI can be used to assess response (Fig. 5.4). The first and most basic approach is to visually assess DWI post-treatment and determine if a high signal remains present within the bowel wall at the site of the tumour. Such a visual assessment has been shown to improve the performance of MRI to discriminate patients with residual tumour from patients with a complete response [23–25]. In the previously described meta-analysis, pooled sensitivity for studies that used DWI for tumour restaging was 84% compared to 50% for standard MRI [22].

The second and so far most often studied approach is to quantitatively measure the tumoural ADC and determine how it changes as a result of treatment. Tumour ADC values typically increase as a result of CRT. This rise in ADC is believed to be caused by a loss of cell membrane integrity, i.e. necrosis. As a late result of CRT, irreparable cell loss will occur, which enlarges the interstitial space leading to increased room for water diffusion and thus increased ADC values [26]. Some groups have reported an initial steep rise in ADC early, i.e. 1–2 weeks, after initiation of treatment, which is believed to be related to tumoural oedema caused by an inflammatory response due to the sudden release of vascular endothelial growth

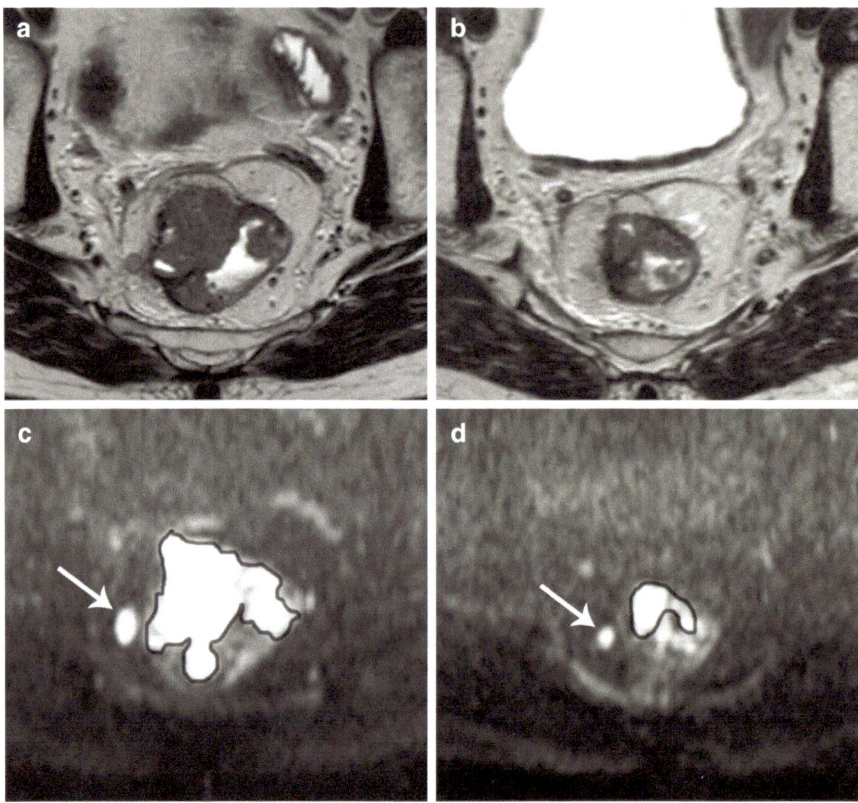

Fig. 5.4 T2-weighted (**a**, **b**) and corresponding high *b*-value diffusion-weighted (**c**, **d**) images of a male patient with a tumour in the mid-rectum imaged before treatment (**a**, **c**) and 8 weeks after completion of chemoradiation (**b**, **d**). The volume of the tumour is delineated on the diffusion-weighted images to calculate the tumour volume and mean tumour ADC. At primary staging, the tumour mass shows clear high signal on DWI; the tumour volume was 10.4 cm^3 and the mean ADC was 1.04×10^{-3} mm^2/s. On the restaging scan, there is still a clear residual tumour mass visible on DWI (**d**). The tumour volume has decreased significantly (to 2.8 cm^3), while the apparent diffusion coefficient increased to 1.32×10^{-3} mm^2/s as a result of treatment. Note the perirectal lymph node (arrow) that also decreased in size as a result of treatment. After surgical resection, histopathology revealed a ypT3N0 tumour

factors [27, 28]. In most published reports, ADC was measured after completion of chemoradiation (typically after an interval of >6 weeks), and the relative change in ADC (ΔADC) compared to the pretreatment ADC was calculated. Results for the use of ADC as a marker to assess response have so far been inconsistent. Some groups showed excellent results for post-treatment and ΔADC to discriminate between good and poor responders, while others found no significant results [27–29]. These conflicting findings are probably also related to the fact that considerable variations in ADC can occur between studies and centres as a result of differences in MR hardware, scan protocols, measurement and post-processing methods. This urges the need for standardisation of protocols before use of ADC as a biomarker of response may be considered in daily practice.

An alternative third approach is to quantify response by calculating the volume of high signal tumour areas on DWI rather than measuring the ADC. Tumour volumes determined on DWI after CRT, as well as the Δvolume, have been shown to be good predictors to differentiate between a complete response and residual tumour with reported accuracies of up to 94%. Moreover, results for DWI volumetry significantly outperformed those of tumour volumetry performed on morphological T2-weighted MRI, as well as results for measuring tumour ADC values [29–31].

5.3.2 DWI for Mesorectal Fascia Assessment After CRT

One report by Park and colleagues specifically evaluated the use of DWI in addition to T2-weighted MRI to predict tumour clearance from the MRF after CRT in a group of 45 patients with clinical suspicion of MRF invasion before treatment. They found that addition of DWI significantly improved the performance of MRI to predict MRF clearance with an AUC of 0.9–0.96 for combined reading of T2W + DWI, compared to AUCs of 0.77–0.85 for T2W-MRI only. Their main explanation was that DWI helps differentiate tumoural invasion from fibrotic and inflammatory stranding into the MRF [32]. Results of this study have so far not been validated by other groups.

5.3.3 DWI for Nodal Restaging

Similar to the results for primary nodal staging, varying results have been reported for nodal restaging with DWI after CRT. Two groups reported significantly higher ADC values for malignant nodes compared to benign/sterilised nodes in the restaging setting. Both studies however also reported that the diagnostic performance for nodal ADC measurements to differentiate between yN0 and yN+ nodes was similar or even poorer than that of routine (size-based) assessment of nodes on T2-weighted MRI [11, 33], indicating that there is no added benefit from a clinical point of view to measure nodal ADCs. Similarly, Ryu et al. reported that addition of DWI did not contribute to improved diagnostic performance compared to T2-weighted MRI to evaluate lymph node eradication (i.e. differentiate between yN0 and yN+ patients) after CRT [34]. Most favourable results so far were published by van Heeswijk et al. who investigated whether the disappearance of nodes on DWI after CRT could be used as a predictor of a node-negative status. They found that when on DWI no high signal nodes remain visible after CRT, this is a 100% sensitive predictor of a ypN0 status, albeit at the cost of a very low specificity of only 14% [35].

5.4 DWI for Follow-Up After Treatment

MRI is not the primary technique of choice for the detection of recurrent disease after curative rectal cancer treatment. Patients with a suspicion of a local recurrence are typically imaged with CT, followed by PET as a second-line modality in case of inconclusive CT findings. MRI is mainly used to assess whether a recurrent tumour is resectable once it has been detected. Whether there is any additional role for DWI in this setting has not yet been investigated. One study evaluated whether it is beneficial to add DWI to standard MRI to detect locally recurrent rectal cancer in patients with a clinical suspicion of a local relapse. They found that, although addition of DWI does not significantly improve overall diagnostic performance, it can improve specificity, i.e. help rule out a recurrence. The authors furthermore suggested that DWI may be beneficial to help detect very early, small, recurrences [36]. Given its sensitivity to detect small tumours, DWI may also prove useful to help detect local tumour regrowths in patients monitored according to a wait-and-see policy. In many of the reported follow-up protocols, these patients regularly undergo MRI and endoscopy with the main aim to detect any recurrence as early as possible so that patients can undergo timely salvage surgery. In this setting, there is some evidence that DWI may aid to detect recurrent disease in an early stage [37].

5.5   DWI as a Prognostic Marker

There is growing interest for the use of ADC as a prognostic imaging biomarker. Low ADC values in primary rectal tumours have been shown to be associated with prognostically unfavourable tumour characteristics such as higher T-stage, the presence of nodal metastases or extranodal tumour deposits, mesorectal fascia invasion and poorer histopathological differentiation grade. Moreover, it has been suggested that pretreatment ADC measurements may hold promise to predict response to neoadjuvant treatment: low pretreatment ADC values have been reported in tumours that respond well to chemoradiotherapy, while tumours with relatively high ADC at primary staging were more prone to respond poorly [27, 38, 39]. It is thought that the high pretreatment ADC in poor responding tumours is related to the presence of necrosis, which decreases the susceptibility of tumours to radiation treatment. Although these results are promising, other study groups have failed to reproduce them or even found contradictory findings [29, 40]. Evidence so far thus remains inconsistent and mainly comes from small-scale, single-centre and retrospective studies with highly varying protocols. Further research is therefore needed to help determine whether ADC can play a meaningful clinical role as a prognostic marker to help further personalise neoadjuvant treatment strategies.

5.6   Pitfalls in Rectal DWI

There are several potential pitfalls that need to be taken into account when assessing the rectum on DWI [41]. First, DWI sequences particularly when acquired using echo planar imaging (EPI) are prone to susceptibility artefacts that tend to occur around gas in the rectal lumen. This may result in distortion of the rectum as well as pile-up of signal projecting over the rectal wall, which—especially when subtle—may be mistaken for tumoural high signal. Such artefacts may be avoided by adapting the sequence acquisition protocol or by reducing the amount of gas in the rectal lumen, for example, by applying endorectal filling or a preparatory microenema. Second, one needs to realise that fibrosis will always appear hypointense on the ADC map, because the high collagen content of fibrosis results in intrinsic short T2 relaxation times with corresponding low or even absent signal. In the absence of a corresponding high signal on high b-value DWI, this low ADC signal should not be mistaken for restricted diffusion caused by tumour. Finally, one always needs to bear in mind that diffusion restriction is not specific for tumour but may also occur as a result of increased viscosity, e.g. in perianal/perirectal abscesses or in tissues with an intrinsically dense cellular structure such as lymphoid tissue. Moreover, in areas of inflammation, high diffusion signal may also be observed to some degree, particularly when scanning with relatively low b-values (b600–b800). As such, diffusion images and ADC maps will always need to be read in conjunction with other anatomical sequences while carefully considering all available clinical information to allow for optimal image interpretation and to draw the right conclusions.

Conclusions

In recent years DWI has increasingly been acknowledged as a valuable adjunct to many oncological MR imaging protocols. In rectal cancer DWI has mainly shown its value in the restaging setting, to help evaluate the response of rectal tumours to neoadjuvant chemoradiotherapy and differentiate viable residual tumour within areas of post-radiation fibrosis. Moreover, the high sensitivity of DWI may aid in detecting small tumours and lymph nodes, although there appears to be little role for DWI to help differentiate between benign/reactive and metastatic lymph nodes. There is some evidence that DWI holds promise as a biomarker to predict treatment outcome and prognosis, although results so far have been inconsistent and will need to be validated by further studies.

References

1. Beets-Tan RGH, Lambregts DMJ, Maas M, Bipat S, Barbaro B, Curvo-Semedo L, Fenlon HM, Gollub MJ, Gourtsoyianni S, Halligan S, Hoeffel C, Kim SH, Laghi A, Maier A, Rafaelsen SR, Stoker J, Taylor SA, Torkzad MR, Blomqvist L. Magnetic resonance imaging for clinical management of rectal cancer: Updated recommendations from the 2016 European Society of Gastrointestinal and Abdominal Radiology (ESGAR) consensus meeting. Eur Radiol. 2018;28:1465–75.
2. Ichikawa T, Erturk SM, Motosugi U, Sou H, Iino H, Araki T, Fujii H. High-B-value diffusion-weighted MRI in colorectal cancer. Am J Roentgenol. 2006;187:181–4.
3. Rao SX, Zeng MS, Chen CZ, Li RC, Zhang SJ, Xu JM, Hou YY. The value of diffusion-weighted imaging in combination with T2-weighted imaging for rectal cancer detection. Eur J Radiol. 2008;65:299–303.
4. Nasu K, Kuroki Y, Minami M. Diffusion-weighted imaging findings of mucinous carcinoma arising in the ano-rectal region: comparison of apparent diffusion coefficient with that of tubular adenocarcinoma. Jpn J Radiol. 2012;30:120–7.
5. Lu ZH, Hu CH, Qian WX, Cao WH. Preoperative diffusion-weighted imaging value of rectal cancer: preoperative T staging and correlations with histological T stage. Clinical imaging. 2016;40:563–8.

6. Feng Q, Yan YQ, Zhu J, Xu JR. T staging of rectal cancer: accuracy of diffusion-weighted imaging compared with T2-weighted imaging on 3.0 tesla MRI. Journal of digestive diseases. 2014;15:188–94.
7. Heijnen LA, Lambregts DM, Mondal D, Martens MH, Riedl RG, Beets GL, Beets-Tan RG. Diffusion-weighted MR imaging in primary rectal cancer staging demonstrates but does not characterise lymph nodes. Eur Radiol. 2013;23:3354–60.
8. Mizukami Y, Ueda S, Mizumoto A, Sasada T, Okumura R, Kohno S, Takabayashi A. Diffusion-weighted magnetic resonance imaging for detecting lymph node metastasis of rectal cancer. World J Surg. 2011;35:895–9.
9. Yasui O, Sato M, Kamada A. Diffusion-weighted imaging in the detection of lymph node metastasis in colorectal cancer. Tohoku J Exp Med. 2009;218:177–83.
10. Cho EY, Kim SH, Yoon JH, Lee Y, Lim YJ, Kim SJ, Baek HJ, Eun CK. Apparent diffusion coefficient for discriminating metastatic from non-metastatic lymph nodes in primary rectal cancer. Eur J Radiol. 2013;82:e662–8.
11. Lambregts DM, Maas M, Riedl RG, Bakers FC, Verwoerd JL, Kessels AG, Lammering G, Boetes C, Beets GL, Beets-Tan RG. Value of ADC measurements for nodal staging after chemoradiation in locally advanced rectal cancer-a per lesion validation study. Eur Radiol. 2011;21:265–73.
12. Lahaye MJ, Engelen SM, Nelemans PJ, Beets GL, van de Velde CJ, van Engelshoven JM, Beets-Tan RG. Imaging for predicting the risk factors--the circumferential resection margin and nodal disease--of local recurrence in rectal cancer: a meta-analysis. Semin Ultrasound CT MR. 2005;26:259–68.
13. Bipat S, Glas AS, Slors FJ, Zwinderman AH, Bossuyt PM, Stoker J. Rectal cancer: local staging and assessment of lymph node involvement with endoluminal US, CT, and MR imaging--a meta-analysis. Radiology. 2004;232:773–83.
14. Kim SH, Yoon JH, Lee Y. Added value of morphologic characteristics on diffusion-weighted images for characterizing lymph nodes in primary rectal cancer. Clinical imaging. 2015;39:1046–51.
15. Kim JH, Beets GL, Kim JH, Kessels AGH, Beets-Tan RGH. High-resolution MR imaging for nodal staging in rectal cancer: are there any criteria in addition to the size? Eur J Radiol. 2004;52:78–83.
16. Brown G, Richards CJ, Bourne MW, Newcombe RG, Radcliffe AG, Dallimore NS, Williams GT. Morphologic predictors of lymph node status in rectal cancer with use of high-spatial-resolution MR imaging with histopathologic comparison. Radiology. 2003;227:371–7.
17. Lezoche G, Baldarelli M, Guerrieri M, Paganini AM, De Sanctis A, Bartolacci S, Lezoche E. A prospective randomized study with a 5-year minimum follow-up evaluation of transanal endoscopic microsurgery versus laparoscopic total mesorectal excision after neoadjuvant therapy. Surg Endosc. 2008;22:352–8.
18. Habr-Gama A, Gama-Rodrigues J, Sao Juliao GP, Proscurshim I, Sabbagh C, Lynn PB, Perez RO. Local recurrence after complete clinical response and watch and wait in rectal cancer after neoadjuvant chemoradiation: impact of salvage therapy on local disease control. Int J Radiat Oncol Biol Phys. 2014;88:822–8.
19. Smith JD, Ruby JA, Goodman KA, Saltz LB, Guillem JG, Weiser MR, Temple LK, Nash GM, Paty PB. Nonoperative management of rectal cancer with complete clinical response after neoadjuvant therapy. Ann Surg. 2012;256:965–72.
20. Martens MH, Maas M, Heijnen LA, Lambregts DM, Leijtens JW, Stassen LP, Breukink SO, Hoff C, Belgers EJ, Melenhorst J, Jansen R, Buijsen J, Hoofwijk TG, Beets-Tan RG, Beets GL. Long-term outcome of an organ preservation program after neoadjuvant treatment for rectal cancer. J Natl Cancer Inst. 2016;108(12)
21. Appelt AL, Ploen J, Harling H, Jensen FS, Jensen LH, Jorgensen JC, Lindebjerg J, Rafaelsen SR, Jakobsen A. High-dose chemoradiotherapy and watchful waiting for distal rectal cancer: a prospective observational study. Lancet Oncol. 2015;16:919–27.
22. van der Paardt MP, Zagers MB, Beets-Tan RG, Stoker J, Bipat S. Patients who undergo preoperative chemoradiotherapy for locally advanced rectal cancer restaged by using diagnostic MR imaging: a systematic review and meta-analysis. Radiology. 2013;269:101–12.

23. Kim SH, Lee JM, Hong SH, Kim GH, Lee JY, Han JK, Choi BI. Locally advanced rectal cancer: added value of diffusion-weighted MR imaging in the evaluation of tumor response to neoadjuvant chemo- and radiation therapy. Radiology. 2009;253:116–25.

24. Lambregts DM, Vandecaveye V, Barbaro B, Bakers FC, Lambrecht M, Maas M, Haustermans K, Valentini V, Beets GL, Beets-Tan RG. Diffusion-weighted MRI for selection of complete responders after chemoradiation for locally advanced rectal cancer: a multicenter study. Ann Surg Oncol. 2011;18:2224–31.

25. Song I, Kim SH, Lee SJ, Choi JY, Kim MJ, Rhim H. Value of diffusion-weighted imaging in the detection of viable tumour after neoadjuvant chemoradiation therapy in patients with locally advanced rectal cancer: comparison with T2-weighted and PET/CT imaging. Br J Radiol. 2012;85:577 86.

26. Seierstad T, Roe K, Olsen DR. Noninvasive monitoring of radiation-induced treatment response using proton magnetic resonance spectroscopy and diffusion-weighted magnetic resonance imaging in a colorectal tumor model. Radiother Oncol. 2007;85:187–94.

27. Sun YS, Zhang XP, Tang L, Ji JF, Gu J, Cai Y, Zhang XY. Locally advanced rectal carcinoma treated with preoperative chemotherapy and radiation therapy: preliminary analysis of diffusion-weighted MR imaging for early detection of tumor histopathologic downstaging. Radiology. 2010;254:170–8.

28. Lambrecht M, Vandecaveye V, De Keyzer F, Roels S, Penninckx F, Van Cutsem E, Claus F, Haustermans K. Value of diffusion-weighted magnetic resonance imaging for prediction and early assessment of response to neoadjuvant radiochemotherapy in rectal cancer: preliminary results. Int J Radiat Oncol Biol Phys. 2012;82:863–70.

29. Curvo-Semedo L, Lambregts DM, Maas M, Thywissen T, Mehsen RT, Lammering G, Beets GL, Caseiro-Alves F, Beets-Tan RG. Rectal cancer: assessment of complete response to preoperative combined radiation therapy with chemotherapy--conventional MR volumetry versus diffusion-weighted MR imaging. Radiology. 2011;260:734–43.

30. Ha HI, Kim AY, Yu CS, Park SH, Ha HK. Locally advanced rectal cancer: diffusion-weighted MR tumour volumetry and the apparent diffusion coefficient for evaluating complete remission after preoperative chemoradiation therapy. Eur Radiol. 2013;23:3345–53.

31. Lambregts DM, Rao SX, Sassen S, Martens MH, Heijnen LA, Buijsen J, Sosef M, Beets GL, Vliegen RA, Beets-Tan RG. MRI and diffusion-weighted MRI volumetry for identification of complete tumor responders after preoperative chemoradiotherapy in patients with rectal cancer: a bi-institutional validation study. Ann Surg. 2015;262:1034–9.

32. Park MJ, Kim SH, Lee SJ, Jang KM, Rhim H. Locally advanced rectal cancer: added value of diffusion-weighted MR imaging for predicting tumor clearance of the mesorectal fascia after neoadjuvant chemotherapy and radiation therapy. Radiology. 2011;260:771–80.

33. Kim SH, Ryu KH, Yoon JH, Lee Y, Paik JH, Kim SJ, Jung HK, Lee KH. Apparent diffusion coefficient for lymph node characterization after chemoradiation therapy for locally advanced rectal cancer. Acta Radiol. 2015;56:1446–53.

34. Ryu KH, Kim SH, Yoon JH, Lee Y, Paik JH, Lim YJ, Lee KH. Diffusion-weighted imaging for evaluating lymph node eradication after neoadjuvant chemoradiation therapy in locally advanced rectal cancer. Acta Radiol. 2016;57:133–41.

35. van Heeswijk MM, Lambregts DM, Palm WM, Hendriks BM, Maas M, Beets GL, Beets-Tan RG. DWI for assessment of rectal cancer nodes after chemoradiotherapy: is the absence of nodes at DWI proof of a negative nodal status? AJR Am J Roentgenol. 2017;208:W79–84.

36. Lambregts DM, Cappendijk VC, Maas M, Beets GL, Beets-Tan RG. Value of MRI and diffusion-weighted MRI for the diagnosis of locally recurrent rectal cancer. Eur Radiol. 2011;21:1250–8.

37. Lambregts DM, Lahaye MJ, Heijnen LA, Martens MH, Maas M, Beets GL, Beets-Tan RG. MRI and diffusion-weighted MRI to diagnose a local tumour regrowth during long-term follow-up of rectal cancer patients treated with organ preservation after chemoradiotherapy. Eur Radiol. 2016;26:2118–25.

38. Intven M, Reerink O, Philippens ME. Diffusion-weighted MRI in locally advanced rectal cancer: pathological response prediction after neo-adjuvant radiochemotherapy. Strahlenther Onkol. 2013;189:117–22.

39. Jung SH, Heo SH, Kim JW, Jeong YY, Shin SS, Soung MG, Kim HR, Kang HK. Predicting response to neoadjuvant chemoradiation therapy in locally advanced rectal cancer: diffusion-weighted 3 Tesla MR imaging. J Magn Reson Imaging. 2012;35:110–6.
40. Monguzzi L, Ippolito D, Bernasconi DP, Trattenero C, Galimberti S, Sironi S. Locally advanced rectal cancer: value of ADC mapping in prediction of tumor response to radiochemotherapy. Eur J Radiol. 2013;82:234–40.
41. Lambregts DMJ, van Heeswijk MM, Delli Pizzi A, van Elderen SGC, Andrade L, Peters NHGM, Kint PAM, Osinga-de Jong M, Bipat S, Ooms R, Lahaye MJ, Maas M, Beets GL, Bakers FCH, Beets-Tan RGH. Diffusion-weighted MRI to assess response to chemoradiotherapy in rectal cancer: main interpretation pitfalls and their use for teaching. Eur Radiol. 2017;27:4445–54.

Anal Canal

6

Sofia Gourtsoyianni and Vicky Goh

6.1 Introduction

A dedicated magnetic resonance imaging (MRI) examination of the lowest part of the gastrointestinal tract, the anus, is performed primarily in two clinical scenarios:

(a) Locoregional staging of anal cancer before and after treatment
(b) Detection/road mapping of perianal fistula disease

The MRI protocols for these two clinical scenarios are different; however, diffusion-weighted imaging (DWI) has been shown to be a very helpful adjunctive sequence providing useful functional information in both.

6.2 Locoregional Staging of Anal Cancer (Baseline)

Squamous cell carcinoma of the anus (SCCA) is a rare malignancy in the general population; however, it is increasingly diagnosed in patients with human immunodeficiency virus (HIV) and chronic inflammatory bowel disease and those having received immunosuppression after transplantation.

Accurate staging of SCCA at presentation, performed according to the 7th UICC/AJCC tumor-node-metastasis (TNM) system [1], provides prognostic information

S. Gourtsoyianni (✉)
Department of Radiology, Guy's and St Thomas' Hospitals NHS Foundation Trust, St Thomas' Hospital, London, UK

V. Goh
Department of Cancer Imaging, School of Biomedical Engineering and Imaging Sciences, King's College London, London, UK
e-mail: vicky.goh@kcl.ac.uk

© Springer International Publishing AG, part of Springer Nature 2019
S. Gourtsoyianni, N. Papanikolaou (eds.), *Diffusion Weighted Imaging of the Gastrointestinal Tract*, https://doi.org/10.1007/978-3-319-92819-7_6

and allows for correct therapeutic planning. In particular, the longest diameter of the primary tumor rather than the extramural extension and the site of involved lymph nodes rather than their actual number are important prognosticators and also comprise major differences from rectal adenocarcinoma staging that might also be encountered extending within the anal canal.

According to the most recent practice guidelines of the European Society for Medical Oncology (ESMO) in collaboration with ESSO and ESTRO [2], MRI is the primary imaging modality of choice for accurate locoregional staging of anal cancer.

Multiplanar, high-resolution T2-weighted sequences, covering the anorectal region, using external phased array body coils, performed on ≥ 1.5 T MRI scanners, allow detailed imaging of the anal sphincter and the surrounding pelvic structures, providing information of local disease extent and nodal involvement, which further assist radiotherapy planning and contouring [2, 3]. In addition short-tau inversion recovery (STIR) sequences are very helpful for identifying fistula tracts [4] present at initial staging or that develop during treatment. The detailed protocol used in our institution for this purpose can be seen in Table 6.1 [5].

The sensitivity of MRI using morphological sequences for the identification of anal SCCs has been reported to approach 90–100%, with high concordance regarding tumor size [6]. On morphological T2W images, SCCA present as intermediate T2W signal intensity mass lesions. MRI unfortunately cannot distinguish low-lying rectal adenocarcinomas from SCCA that may extend upward involving the rectum as the two cancer types share similar morphological imaging characteristics. Qualitative assessment of DWI images of the two entities is also not helpful as both entities are hypercellular and demonstrate a degree of restricted diffusion (Fig. 6.1).

The incidence of involved regional lymph nodes increases with primary tumor size. Lymph node metastases may be present in 25% of cases even with superficial, $\leq$T2 stage tumors, at initial staging [7]. Nodal staging relies on the distance of

Table 6.1 MRI protocol for locoregional staging of SCCA

MRI	Sequence	Acquisition parameters
Pelvis	T2 axial TSE	TR/TE, 6500/133 ms; NEX 1, ST 6 mm
	T1 axial SE	TR/TE, 495/20 ms; NEX 1, ST 6 mm
	T2 sagittal TSE	TR, 6500/135 ms; NEX 1, ST 5 mm
	Diffusion axial:	TR, 3500/97 ms; NEX 6, ST 6 mm
	T2-W SS-EPI; three directions;	
	$b = 0, 1400$ s/mm^2	
Tumor	T2 axial[a]	TR/TE, 6500/137 ms; NEX 4, ST 4 mm
	T2 coronal[b]	
	TSE	
	STIR axial[a]	TR/TE/TI, 5750/25/150 ms; NEX 1,
	STIR coronal[b]	ST 4 mm
	Diffusion axial[a]:	TR/TE, 3000/83; NEX 6, ST 6 mm
	T2-W SS-EPI; three directions;	
	$b = 0, 50, 100, 400, 800$ s/mm^2	

NEX number of excitations, *ST* slice thickness
[a]Perpendicular and [b]parallel to long axis of anal canal

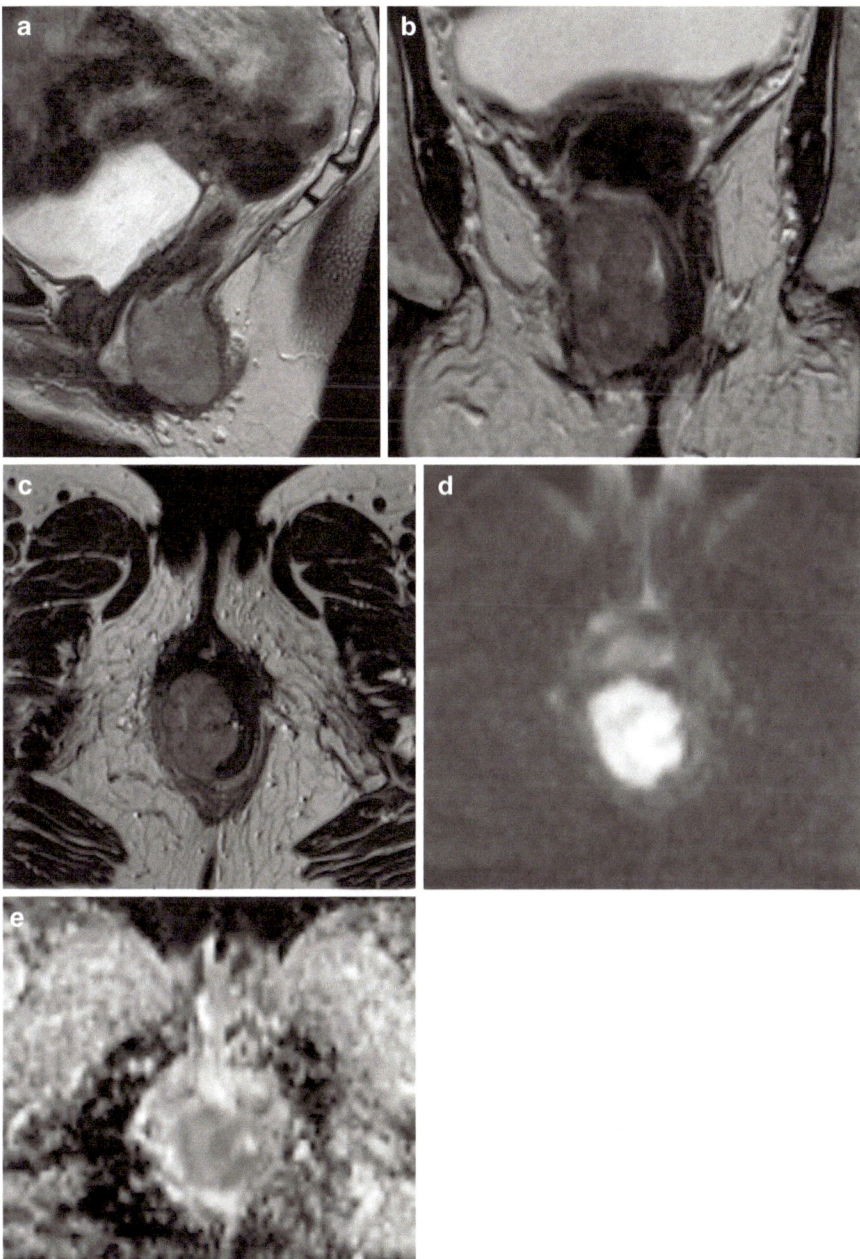

Fig. 6.1 A 63-year-old female patient presented with a SCCA and underwent baseline MRI scan for locoregional staging purposes. High-resolution T2-TSE images on sagittal (**a**), coronal (**b**), and axial (**c**) plane demonstrated a 4.5 cm in longest diameter mass involving the right anal sphincter complex with no visible locoregional lymph nodes rendering this a T2N0. In addition T2-W SS-EPI DWI sequence with three directions, $b = 0, 50, 100, 400, 800$ s/mm^2, was obtained. The mass demonstrates high signal intensity on b800 image (**d**) and low signal intensity on corresponding ADC map (**e**) in keeping with restricted diffusion

nodes from the primary tumor site rather than on the number of involved nodes. MRI, especially with the implementation of DWI, is helpful in identifying pelvic lymph nodes. As with rectal adenocarcinoma, it has been found that almost half of all involved lymph nodes have a diameter <5 mm. Short-axis threshold values of 8 mm, 5 mm, and 10 mm have been suggested for pelvic, mesorectal, and inguinal lymph nodes, respectively [8]. Involved lymph nodes have been reported as having signal intensity similar to or the same as that of the primary tumor [3].

A recently published retrospective study investigated in a cohort of 58 patients both qualitative and quantitative DWI parameters, namely, signal intensity and ADC values of primary lesion and lymph nodes [9], and found that neoplastic nodes showed similar signal intensity and ADC value to the anal cancer. The signal of both primary anal cancer and neoplastic lymph nodes was hyperintense on b 800 s/mm^2, highest b value obtained, and hypointense on corresponding ADC map in keeping with restricted diffusion and were found to have overlapping ADC values (Fig. 6.2).

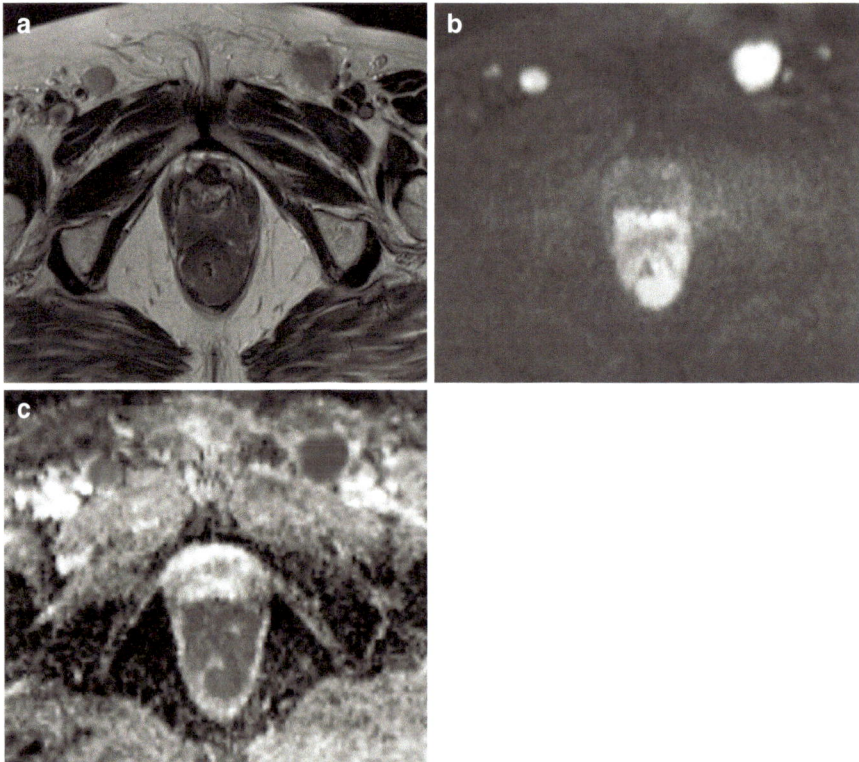

Fig. 6.2 A 77-year-old female patient presented with SCCA extending in the lower rectum (**a**). b800 image (**b**) and corresponding ADC map (**c**) certainly add to the radiologist's confidence in delineating the primary tumor which demonstrates restricted diffusion, same as would have been in the case of a low-lying rectal adenocarcinoma, in addition to providing an excellent road map of disease extent with prominent bilateral inguinal lymph nodes also depicted

A promising prospective study design has been published by Jones et al. [10] with ongoing recruitment incorporating DWI in their imaging protocol. DWI is to be performed at four different time points, namely, at baseline, 2nd and 4th week of treatment and 6–8 weeks posttreatment, respectively, and four *b* values are to be used (0, 400, 800, 1200) at a 3 T scanner. The aim of the study is to test the ability of DWI to serve as an imaging biomarker, to investigate potential predictive value of quantitative ADC measurements performed to assess response and potentially tailor radiotherapy dose according to individual patient's tumor response.

A protocol for a feasibility study (*n* = 11) has also been published investigating brachytherapy to boost EBRT results introducing an MR/CT image adapted brachytherapy (MR/CT-IABT) protocol. The MRI protocol includes DWI sequence as authors believe that more adequate clinical target volume and biological information may be provided [11].

This is supported by recent data where it has been found that sequence selection affects primary anorectal cancer volume measurements [12] and subsequently the T stage. DWI yielded higher interobserver agreement and greater tumor delineation confidence compared to standard of care high-resolution T2-weighted sequences providing lower gross tumor volume (GTV) and maximum tumor diameter (MTD) [13].

6.3 Locoregional Staging of Anal Cancer After Treatment

First-line treatment for SCCA is definitive chemoradiation therapy (CRT), comprising external-beam radiotherapy (EBRT) with a dose ranging between 45 and 50 Gy, combined with mitomycin-C and infusional 5-FU chemotherapy. In 35% of cases, usually patients presenting with advanced T3/T4 stage tumors, locoregional and/or metastatic relapse may occur [4]. A boost by means of intensity-modulated radiotherapy (IMRT) and brachytherapy (BT) is recommended for good local control in high-risk patients.

Koh et al. [14] were the first to report the range of post-CRT MRI appearances of anal cancers using T2-weighted and short-tau inversion recovery (STIR) imaging before and after chemoradiation. Tumor response was assessed by recording change in tumor size, signal intensity, distortion of anal canal/sphincter complex, infiltration of adjacent structures, and nodal disease immediately after chemoradiation, every 6 months for the first year and then yearly. Responders with long disease remission demonstrated the greatest size involution of MR signal abnormality in the tumor area at 6 months after treatment. This had led to the belief that MRI assessment is more beneficial in demonstrating changes post-CRT if performed at a later stage than that recommended for rectal adenocarcinoma (6–8 weeks post-CRT).

Goh et al. [15] applied RECIST criteria for categorizing patients into responders and nonresponders (*n* = 35) at 6–8 weeks post-CRT and found no difference between disease-free and relapsed patients among the two groups, supporting that RECIST response based on MRI at 6–8 weeks is not a relevant end point to explore phase II trials, novel treatments, and CRT combinations.

Kochhar et al. [16] have proposed an MRI tumor regression grade (TRG) scoring system for assessing response post-chemoradiotherapy akin to the MRI tumor regression grading for rectal cancer; this divides response into five groups with a TRG of 5 indicating no response. This has yet to be validated.

Unfortunately there is yet no established consensus on the optimal timing for documenting SCCA response post-CRT MRI examination. Complete response (CR) of SCCA has been reported to occur at around 26 weeks [17], while 50% of local recurrences occurring within the first 2 years posttreatment are located around the primary site of disease or present as pelvic/inguinal lymph nodes. MRI is thought to serve as a useful complement to clinical evaluation and provides a more comprehensive assessment of therapeutic response after CRT, especially for treatment-related changes such as fibrosis and presence of lymph nodes, not accessible with endoanal ultrasound (EAUS).

The same MRI protocol (Table 6.1) is performed for therapy assessment. The appearance of T2W hypointense signal at the site of primary tumor is consistent with fibrosis, a morphological sign of response; however, morphological MRI sequences, namely, high-resolution T2W images, cannot exclude residual neoplastic foci within dense fibrosis, the same problem encountered with CRT treated locally advanced rectal adenocarcinomas. Addition of DWI to T2W imaging has been shown to improve the prediction of tumor clearance in the mesorectal fascia (MRF) after neoadjuvant CRT compared with T2-weighted imaging alone in patients with locally advanced rectal cancer, who develop post-radiotherapy fibrosis [18]. Interobserver agreement of confidence levels has also been reported to be better for the combined set of DWI and T2-weighted images.

As far as involved lymph nodes are concerned, these tend to demonstrate same changes post therapy as the primary tumor. Morphological criteria used for lymph node characterization have been reported to work better post-chemoradiotherapy for rectal cancer [19], recognizing MRI as a useful tool for assessment of treatment. In addition introduction of DWI for lymph detection and characterization has so far succeeded in improving identification of small pelvic lymph nodes, as is the case of mesorectal lymph nodes especially when morphological T2W images are fused with highest b value obtained DWI images [20].

In a recently published study [10], it was found that the signal intensity of responding residual anal tumor after treatment was less hyperintense on b800 and less hypointense on ADC map, demonstrating less restricted diffusion, while ADC values were higher. The same applied in responders regarding lymph nodes on DWI. Nonresponders did not show any significant differences in diffusion restriction or ADC values between pre- and posttreatment.

As anal cancer patients tend to undergo long-term follow-up and surveillance, MRI may also assist in the early detection of disease relapse. Diffusion-weighted imaging appears to have an emerging role for differentiating suspected small residual/recurrent tumor from treatment-related changes [21]; however, no dedicated studies have been published so far.

6.4   Perianal Fistula Disease Detection/Road Mapping

A second potential use of DWI in the anal canal is the detection and evaluation of perianal fistula activity. However few studies have been published on this.

Cavusoglu et al. [22] reported that DWI using b values of 0 and 1000 s/mm^2 has a significant added value, providing both higher sensitivity and specificity, compared to fat-suppressed T2-weighted fast spin-echo imaging alone in the diagnosis of perianal fistula. In a recent review of MRI for perianal fistula [23], it was pointed out that although some authors propose a potential routine role for diffusion-weighted imaging [24] and in particular for the diagnosis of abscess-complicating fistula-in-ano that might obviate the requirement for contrast administration, there are downsides to DWI assessment of the perianal region. In particular, there are difficulties in interpretation resulting from artifacts close to air–soft tissue interfaces, which can degrade image quality [25].

Yoshizako et al. [26] included DWI, b values of 0 and 1000 s/mm^2, in imaging of a small cohort of 24 patients with clinically suspected perianal fistula treated conservatively with antibiotics. ADC measurements of the lesions, classified into two groups based on the need for surgery and surgical findings, namely, positive inflammation activity (PIA) and negative inflammation activity (NIA) groups, were performed.

The ADC of the PIA group was significantly lower than that of the NIA group, demonstrating promise for quantitative use of DWI for the evaluation of perianal fistula activity.

In addition feasibility and reproducibility of diffusion tensor imaging (DTI), a specific type of modeling of the DWI datasets, of the anal canal have been investigated at 3 T. Goh et al. [27] performed DTI in 25 men with no anal canal disease symptomatology. Fractional anisotropy (FA), relative anisotropy (RA), and apparent diffusion coefficient (ADC) were determined for the epithelial/subepithelial layer, internal sphincter, external sphincter, and puborectalis with a good overall intra- and inter-rater agreement and test-retest reproducibility noted. However, the method has not been tested on patients with perianal fistula disease history so far according to our knowledge.

Conclusion

Diffusion-weighted imaging is a helpful adjunctive sequence to the MRI examination protocol of the anal canal providing useful functional information for both locoregional staging of anal cancer and detection/road mapping of perianal fistula disease. Studies in anal cancer are underway investigating its role as an imaging biomarker to potentially tailor radiotherapy dose according to individual patient's tumor/tumor response. Diffusion-weighted imaging has improved interobserver agreement and provided a higher level of confidence in tumor delineation compared to standard of care high-resolution T2-weighted sequences. Diffusion-weighted imaging has also shown promise in detection of residual cancer as well as differentiating suspected recurrent tumor from treatment-related changes.

References

1. Edge SB, Byrd DR, Compton CC, Fritz AG, Greene FL, Trotti A, editors. AJCC cancer staging manual. 7th ed. New York, NY: Springer; 2010.
2. Glynne-Jones R, Nilsson PJ, Aschele C, et al. Anal cancer: ESMO-ESSO-ESTRO clinical practice guidelines for diagnosis, treatment and follow-up. Radiother Oncol. 2014;111(3):330–9.
3. Roach SC, Hulse PA, Moulding FJ, Wilson R, Carrington BM. Magnetic resonance imaging of anal cancer. Clin Radiol. 2005;60(10):1111–9.
4. Kochhar R, Plumb AA, Carrington BM, Saunders M. Imaging of anal carcinoma. AJR Am J Roentgenol. 2012;199(3):W335–44.
5. Gourtsoyianni S, Goh V. MRI of anal cancer: assessing response to definitive chemoradiotherapy. Abdom Imaging. 2014;39(1):2–17.
6. Otto SD, Lee L, Buhr HJ, Frericks B, Höcht S, Kroesen AJ. Staging anal cancer: prospective comparison of transanal endoscopic ultrasound and magnetic resonance imaging. J Gastrointest Surg. 2009;13(7):1292–8.
7. Raghunathan G, Mortele KJ. Magnetic resonance imaging of anorectal neoplasms. Clin Gastroenterol Hepatol. 2009;7(4):379–88.
8. Tonolini M, Bianco R. MRI and CT of anal carcinoma: a pictorial review. Insights Imaging. 2013;4(1):53–62.
9. Reginelli A, Granata V, Fusco R, et al. Diagnostic performance of magnetic resonance imaging and 3D endoanal ultrasound in detection, staging and assessment post treatment, in anal cancer. Oncotarget. 2017;8(14):22980–90.
10. Jones M, Hruby G, Stanwell P, et al. Multiparametric MRI as an outcome predictor for anal canal cancer managed with chemoradiotherapy. BMC Cancer. 2015;15:281.
11. Tagliaferri L, Manfrida S, Barbaro B, et al. MITHRA - multiparametric MR/CT image adapted brachytherapy (MR/CT-IABT) in anal canal cancer: a feasibility study. J Contemp Brachytherapy. 2015;7(5):336–45.
12. Regini F, Gourtsoyianni S, Cardoso De Melo R, et al. Rectaltumour volume (GTV)delineation using T2-weighted and diffusion-weighted MRI: implications for radiotherapy planning. Eur J Radiol. 2014;83(5):768–72.
13. Prezzi D, Mandegaran R, Gourtsoyianni S, Owczarczyk K, Gaya A, Glynne-Jones R, Goh V. The impact of MRI sequence on tumour staging and gross tumour volume delineation in squamous cell carcinoma of the anal canal. Eur Radiol. 2017. https://doi.org/10.1007/s00330-017-5133-0.
14. Koh DM, Dzik-Jurasz A, O'Neill B, Tait D, Husband JE, Brown G. Pelvic phased-array MR imaging of anal carcinoma before and after chemoradiation. Br J Radiol. 2008;81(962):91–8.
15. Goh V, Gollub FK, Liaw J, et al. Magnetic resonance imaging assessment of squamous cell carcinoma of the anal canal before and after chemoradiation: can MRI predict for eventual clinical outcome? Int J Radiat Oncol Biol Phys. 2010;78(3):715–21.
16. Kochhar R, Renehan AG, Mullan D, Chakrabarty B, Saunders MP, Carrington BM. The assessment of local response using magnetic resonance imaging at 3- and6-month post chemoradiotherapy in patients with anal cancer. Eur Radiol. 2017;27(2):607–17.
17. Glynne-Jones R, James R, Meadows H, et al. ACT II Study Group. Optimum time to assess complete clinical response (CR) following chemoradiation (CRT) using mitomycin (MMC) or cisplatin (CisP), with or without maintenance CisP/5FU in squamous cell carcinoma of the anus: results of ACT II. J Clin Oncol. 2012;30(15_suppl):4004.
18. Park MJ, Kim SH, Lee SJ, Jang KM, Rhim H. Locally advanced rectal cancer: added value of diffusion-weighted MR imaging for predicting tumor clearance of the mesorectal fascia after neoadjuvant chemotherapy and radiation therapy. Radiology. 2011;260(3):771–80.
19. Koh DM, Chau I, Tait D, Wotherspoon A, Cunningham D, Brown G. Evaluating mesorectal lymph nodes in rectal cancer before and after neoadjuvant chemoradiation using thin-section T2-weighted magnetic resonance imaging. Int J Radiat Oncol Biol Phys. 2008;71(2):456–61.

20. Mir N, Sohaib SA, Collins D, Koh DM. Fusion of high b-value diffusion-weighted and T2-weighted MR images improves identification of lymph nodes in the pelvis. J Med Imaging Radiat Oncol. 2010;54(4):358–64.
21. Renehan AG, Saunders MP, Schofield PF, O'Dwyer ST. Patterns of local disease failure and outcome after salvage surgery in patients with anal cancer. Br J Surg. 2005;92(5):605–14.
22. Cavusoglu M, Duran S, SözmenCılız D, Tufan G, HatipogluÇetin HG, Ozsoy A, Sakman B. Added value of diffusion-weighted magnetic resonance imaging for the diagnosis of perianal fistula. Diagn Interv Imaging. 2017;98(5):401–8.
23. Tolan DJ. Magnetic resonance imaging for perianal fistula. Semin Ultrasound CT MR. 2016;37(4):313–22.
24. Hori M, Oto A, Orrin S, Suzuki K, Baron RL. Diffusion-weighted MRI: a new tool for the diagnosis of fistula in ano. J Magn Reson Imaging. 2009;30(5):1021–6.
25. Dohan A, Eveno C, Oprea R, Pautrat K, Placé V, Pocard M, Hoeffel C, Boudiaf M, Soyer P. Diffusion-weighted MR imaging for the diagnosis of abscess complicating fistula-in-ano: preliminary experience. Eur Radiol. 2014;24(11):2906–15.
26. Yoshizako T, Wada A, Takahara T, et al. Diffusion-weighted MRI for evaluating perianal fistula activity: feasibility study. Eur J Radiol. 2012;81(9):2049–53.